EXPERIMENTAL AND QUASI-EXPERIMENTAL DESIGNS FOR RESEARCH

EXPERIMENTAL AND QUASI-EXPERIMENTAL DESIGNS FOR RESEARCH

DONALD T. CAMPBELL
Northwestern University

JULIAN C. STANLEY
Johns Hopkins University

RAND McNALLY COLLEGE PUBLISHING COMPANY
CHICAGO

Current printing (last digit)
15 14 13 12 11

Reprinted from *Handbook of Research on Teaching*

Printed in U.S.A.
Library of Congress Catalogue Card Number 66:25213

Preface

This survey originally appeared in N. L. Gage (Editor), *Handbook of Research on Teaching,* published by Rand McNally & Company in 1963, under the longer title "Experimental and Quasi-Experimental Designs for Research on Teaching." As a result, the introductory pages and many of the illustrations come from educational research. But as a study of the references will indicate, the survey draws from the social sciences in general, and the methodological recommendations are correspondingly broadly appropriate.

For the convenience of the user we have added a table of contents, a list of supplementary references, a name index and a subject index.

DONALD T. CAMPBELL
JULIAN C. STANLEY

1966

Contents

Experimental and Quasi-Experimental Designs for Research[1]

DONALD T. CAMPBELL
Northwestern University

JULIAN C. STANLEY
Johns Hopkins University

In this chapter we shall examine the validity of 16 experimental designs against 12 common threats to valid inference. By experiment we refer to that portion of research in which variables are manipulated and their effects upon other variables observed. It is well to distinguish the particular role of this chapter. It is *not* a chapter on experimental design in the Fisher (1925, 1935) tradition, in which an experimenter having complete mastery can schedule treatments and measurements for optimal statistical efficiency, with complexity of design emerging only from that goal of efficiency. Insofar as the designs discussed in the present chapter become complex, it is because of the intransigency of the environment: because, that is, of the experimenter's lack of complete control. While contact is made with the Fisher tradition at several points, the exposition of that tradition is appropriately left to full-length presentations, such as the books by Brownlee (1960), Cox (1958), Edwards (1960), Ferguson (1959), Johnson (1949), Johnson and Jackson (1959), Lindquist (1953), McNemar (1962), and Winer (1962). (Also see Stanley, 1957b.)

PROBLEM AND BACKGROUND

McCall as a Model

In 1923, W. A. McCall published a book entitled *How to Experiment in Education*. The present chapter aspires to achieve an up-to-date representation of the interests and considerations of that book, and for this reason will begin with an appreciation of it. In his preface McCall said: "There are excellent books and courses of instruction dealing with the statistical manipulation of experimental data, but there is little help to be found on the methods of securing adequate and proper data to which to apply statistical procedure." This sentence remains true enough today to serve as the leitmotif of this presentation also. While the impact of the Fisher tradition has remedied the situation in some fundamental ways, its most conspicuous effect seems to have been to

[1] The preparation of this chapter has been supported by Northwestern University's Psychology-Education Project, sponsored by the Carnegie Corporation. Keith N. Clayton and Paul C. Rosenblatt have assisted in its preparation.

elaborate statistical analysis rather than to aid in securing "adequate and proper data."

Probably because of its practical and common-sense orientation, and its lack of pretension to a more fundamental contribution, McCall's book is an undervalued classic. At the time it appeared, two years before the first edition of Fisher's *Statistical Methods for Research Workers* (1925), there was nothing of comparable excellence in either agriculture or psychology. It anticipated the orthodox methodologies of these other fields on several fundamental points. Perhaps Fisher's most fundamental contribution has been the concept of achieving pre-experimental equation of groups through randomization. This concept, and with it the rejection of the concept of achieving equation through matching (as intuitively appealing and misleading as that is) has been difficult for educational researchers to accept. In 1923, McCall had the fundamental qualitative understanding. He gave, as his first method of establishing comparable groups, "groups equated by chance." "Just as representativeness can be secured by the method of chance, . . . so equivalence may be secured by chance, provided the number of subjects to be used is sufficiently numerous" (p. 41). On another point Fisher was also anticipated. Under the term "rotation experiment," the Latin-square design was introduced, and, indeed, had been used as early as 1916 by Thorndike, McCall, and Chapman (1916), in both 5×5 and 2×2 forms, i.e., some 10 years before Fisher (1926) incorporated it systematically into his scheme of experimental design, with randomization.[2]

McCall's mode of using the "rotation experiment" serves well to denote the emphasis of his book and the present chapter. The rotation experiment is introduced not for reasons of efficiency but rather to achieve some degree of control where random assignment to equivalent groups is not possible. In a similar vein, this chapter will examine the imperfections of numerous experimental schedules and will nonetheless advocate their utilization in those settings where better experimental designs are not feasible. In this sense, a majority of the designs discussed, including the unrandomized "rotation experiment," are designated as *quasi*-experimental designs.

Disillusionment with Experimentation in Education

This chapter is committed to the experiment: as the only means for settling disputes regarding educational practice, as the only way of verifying educational improvements, and as the only way of establishing a cumulative tradition in which improvements can be introduced without the danger of a faddish discard of old wisdom in favor of inferior novelties. Yet in our strong advocacy of experimentation, we must not imply that our emphasis is new. As the existence of McCall's book makes clear, a wave of enthusiasm for experimentation dominated the field of education in the Thorndike era, perhaps reaching its apex in the 1920s. And this enthusiasm gave way to apathy and rejection, and to the adoption of new psychologies unamenable to experimental verification. Good and Scates (1954, pp. 716–721) have documented a wave of pessimism, dating back to perhaps 1935, and have cited even that staunch advocate of experimentation, Monroe (1938), as saying "the direct contributions from controlled experimentation have been disappointing." Further, it can be noted that the defections from experimentation to essay writing, often accompanied by conversion from a Thorndikian behaviorism to Gestalt psychology or psychoanalysis, have frequently occurred in persons well trained in the experimental tradition.

To avoid a recurrence of this disillusionment, we must be aware of certain sources of the previous reaction and try to avoid the false anticipations which led to it. Several aspects may be noted. First, the claims made for the rate and degree of progress which would result from experiment were grandi-

[2] Kendall and Buckland (1957) say that the Latin square was invented by the mathematician Euler in 1782. Thorndike, Chapman, and McCall do not use this term.

osely overoptimistic and were accompanied by an unjustified depreciation of nonexperimental wisdom. The initial advocates assumed that progress in the technology of teaching had been slow *just because* scientific method had not been applied: they assumed traditional practice was incompetent, just because it had not been produced by experimentation. When, in fact, experiments often proved to be tedious, equivocal, of undependable replicability, and to confirm prescientific wisdom, the overoptimistic grounds upon which experimentation had been justified were undercut, and a disillusioned rejection or neglect took place.

This disillusionment was shared by both observer and participant in experimentation. For the experimenters, a personal avoidance-conditioning to experimentation can be noted. For the usual highly motivated researcher the nonconfirmation of a cherished hypothesis is actively painful. As a biological and psychological animal, the experimenter is subject to laws of learning which lead him inevitably to associate this pain with the contiguous stimuli and events. These stimuli are apt to be the experimental process itself, more vividly and directly than the "true" source of frustration, i.e., the inadequate theory. This can lead, perhaps unconsciously, to the avoidance or rejection of the experimental process. If, as seems likely, the ecology of our science is one in which there are available many more wrong responses than correct ones, we may anticipate that most experiments will be disappointing. We must somehow inoculate young experimenters against this effect, and in general must justify experimentation on more pessimistic grounds—not as a panacea, but rather as the only available route to cumulative progress. We must instill in our students the expectation of tedium and disappointment and the duty of thorough persistence, by now so well achieved in the biological and physical sciences. We must expand our students' vow of poverty to include not only the willingness to accept poverty of finances, but also a poverty of experimental results.

More specifically, we must increase our time perspective, and recognize that continuous, multiple experimentation is more typical of science than once-and-for-all definitive experiments. The experiments we do today, if successful, will need replication and cross-validation at other times under other conditions before they can become an established part of science, before they can be theoretically interpreted with confidence. Further, even though we recognize experimentation as the basic language of proof, as the only decision court for disagreement between rival theories, we should not expect that "crucial experiments" which pit opposing theories will be likely to have clear-cut outcomes. When one finds, for example, that competent observers advocate strongly divergent points of view, it seems likely on a priori grounds that both have observed something valid about the natural situation, and that both represent a part of the truth. The stronger the controversy, the more likely this is. Thus we might expect in such cases an experimental outcome with mixed results, or with the balance of truth varying subtly from experiment to experiment. The more mature focus—and one which experimental psychology has in large part achieved (e.g., Underwood, 1957b)—avoids crucial experiments and instead studies dimensional relationships and interactions along many degrees of the experimental variables.

Not to be overlooked, either, are the greatly improved statistical procedures that quite recently have filtered slowly into psychology and education. During the period of its greatest activity, educational experimentation proceeded ineffectively with blunt tools. McCall (1923) and his contemporaries did one-variable-at-a-time research. For the enormous complexities of the human learning situation, this proved too limiting. We now know how important various contingencies—dependencies upon joint "action" of two or more experimental variables—can be. Stanley (1957a, 1960, 1961b, 1961c, 1962), Stanley and Wiley (1962), and others have stressed the assessment of such interactions.

Experiments may be multivariate in either or both of two senses. More than one "independent" variable (sex, school grade, method of teaching arithmetic, style of printing type, size of printing type, etc.) may be incorporated into the design and/or more than one "dependent" variable (number of errors, speed, number right, various tests, etc.) may be employed. Fisher's procedures are multivariate in the first sense, univariate in the second. Mathematical statisticians, e.g., Roy and Gnanadesikan (1959), are working toward designs and analyses that unify the two types of multivariate designs. Perhaps by being alert to these, educational researchers can reduce the usually great lag between the introduction of a statistical procedure into the technical literature and its utilization in substantive investigations.

Undoubtedly, training educational researchers more thoroughly in *modern* experimental statistics should help raise the quality of educational experimentation.

Evolutionary Perspective on Cumulative Wisdom and Science

Underlying the comments of the previous paragraphs, and much of what follows, is an evolutionary perspective on knowledge (Campbell, 1959), in which applied practice and scientific knowledge are seen as the resultant of a cumulation of selectively retained tentatives, remaining from the hosts that have been weeded out by experience. Such a perspective leads to a considerable respect for tradition in teaching practice. If, indeed, across the centuries many different approaches have been tried, if some approaches have worked better than others, and if those which worked better have therefore, to some extent, been more persistently practiced by their originators, or imitated by others, or taught to apprentices, then the customs which have emerged may represent a valuable and tested subset of all possible practices.

But the selective, cutting edge of this process of evolution is very imprecise in the natural setting. The conditions of observation, both physical and psychological, are far from optimal. What survives or is retained is determined to a large extent by pure chance. Experimentation enters at this point as the means of sharpening the relevance of the testing, probing, selection process. Experimentation thus is not in itself viewed as a source of ideas necessarily contradictory to traditional wisdom. It is rather a refining process superimposed upon the probably valuable cumulations of wise practice. Advocacy of an experimental science of education thus does not imply adopting a position incompatible with traditional wisdom.

Some readers may feel a suspicion that the analogy with Darwin's evolutionary scheme becomes complicated by specifically human factors. School principal John Doe, when confronted with the necessity for deciding whether to adopt a revised textbook or retain the unrevised version longer, probably chooses on the basis of scanty knowledge. Many considerations besides sheer efficiency of teaching and learning enter his mind. The principal can be right in two ways: keep the old book when it is as good as or better than the revised one, or adopt the revised book when it is superior to the unrevised edition. Similarly, he can be wrong in two ways: keep the old book when the new one is better, or adopt the new book when it is no better than the old one.

"Costs" of several kinds might be estimated roughly for each of the two erroneous choices: (1) financial and energy-expenditure cost; (2) cost to the principal in complaints from teachers, parents, and schoolboard members; (3) cost to teachers, pupils, and society because of poorer instruction. These costs in terms of money, energy, confusion, reduced learning, and personal threat must be weighed against the probability that each will occur and also the probability that the error itself will be detected. If the principal makes his decision without suitable research evidence concerning Cost 3 (poorer instruction), he is likely to overemphasize Costs 1 and 2. The cards seem stacked in favor of a conservative approach—that is, retaining the old book for another year. We can, however, try to cast an experiment with the two books into a decision-theory mold (Chernoff & Moses, 1959) and reach a decision that takes the various costs and probabilities into consideration explicitly. How nearly the careful deliberations of an excellent educational administrator approximate this decision-theory model is an important problem which should be studied.

Factors Jeopardizing Internal and External Validity

In the next few sections of this chapter we spell out 12 factors jeopardizing the validity of various experimental designs.[3] Each factor will receive its main exposition in the context of those designs for which it is a particular problem, and 10 of the 16 designs will be presented before the list is complete. For purposes of perspective, however, it seems well to provide a list of these factors and a general guide to Tables 1, 2, and 3, which partially summarize the discussion. Fundamental to this listing is a distinction between *internal validity* and *external validity*. *Internal validity* is the basic minimum without which any experiment is uninterpretable: Did in fact the experimental treatments make a difference in this specific experimental instance? *External validity* asks the question of *generalizability:* To what populations, settings, treatment variables, and measurement variables can this effect be generalized? Both types of criteria are obviously important, even though they are frequently at odds in that features increasing one may jeopardize the other. While *internal validity* is the *sine qua non,* and while the question of *external validity,* like the question of inductive inference, is never completely answerable, the selection of designs strong in both types of validity is obviously our ideal. This is particularly the case for research on teaching, in which generalization to applied settings of known character is the desideratum. Both the distinctions and the relations between these two classes of validity considerations will be made more explicit as they are illustrated in the discussions of specific designs.

Relevant to *internal validity,* eight different classes of extraneous variables will be presented; these variables, if not controlled in the experimental design, might produce effects confounded with the effect of the experimental stimulus. They represent the effects of:

1. *History,* the specific events occurring between the first and second measurement in addition to the experimental variable.
2. *Maturation,* processes within the respondents operating as a function of the passage of time per se (not specific to the particular events), including growing older, growing hungrier, growing more tired, and the like.
3. *Testing,* the effects of taking a test upon the scores of a second testing.
4. *Instrumentation,* in which changes in the calibration of a measuring instrument or changes in the observers or scorers used may produce changes in the obtained measurements.
5. *Statistical regression,* operating where groups have been selected on the basis of their extreme scores.
6. Biases resulting in differential *selection* of respondents for the comparison groups.
7. *Experimental mortality,* or differential loss of respondents from the comparison groups.
8. *Selection-maturation interaction,* etc., which in certain of the multiple-group quasi-experimental designs, such as Design 10, is confounded with, i.e., might be mistaken for, the effect of the experimental variable.

The factors jeopardizing *external validity* or *representativeness* which will be discussed are:

9. The *reactive* or *interaction effect* of *testing,* in which a pretest might increase or

[3] Much of this presentation is based upon Campbell (1957). Specific citations to this source will, in general, not be made.

decrease the respondent's sensitivity or responsiveness to the experimental variable and thus make the results obtained for a pretested population unrepresentative of the effects of the experimental variable for the unpretested universe from which the experimental respondents were selected.

10. The *interaction* effects of *selection* biases and the *experimental variable*.

11. *Reactive effects of experimental arrangements,* which would preclude generalization about the effect of the experimental variable upon persons being exposed to it in nonexperimental settings.

12. *Multiple-treatment interference,* likely to occur whenever multiple treatments are applied to the same respondents, because the effects of prior treatments are not usually erasable. This is a particular problem for one-group designs of type 8 or 9.

In presenting the experimental designs, a uniform code and graphic presentation will be employed to epitomize most, if not all, of their distinctive features. An X will represent the exposure of a group to an experimental variable or event, the effects of which are to be measured; O will refer to some process of observation or measurement; the Xs and Os in a given row are applied to the same specific persons. The left-to-right dimension indicates the temporal order, and Xs and Os vertical to one another are simultaneous. To make certain important distinctions, as between Designs 2 and 6, or between Designs 4 and 10, a symbol R, indicating random assignment to separate treatment groups, is necessary. This randomization is conceived to be a process occurring at a specific time, and is the all-purpose procedure for achieving pretreatment equality of groups, within known statistical limits. Along with this goes another graphic convention, in that parallel rows unseparated by dashes represent comparison groups equated by randomization, while those separated by a dashed line represent comparison groups not equated by random assignment. A symbol for matching as a process for the pretreatment equating of comparison groups has not been used, because the value of this process has been greatly oversold and it is more often a source of mistaken inference than a help to valid inference. (See discussion of Design 10, and the final section on correlational designs, below.) A symbol M for materials has been used in a specific way in Design 9.

THREE PRE-EXPERIMENTAL DESIGNS

1. The One-Shot Case Study

Much research in education today conforms to a design in which a single group is studied only once, subsequent to some agent or treatment presumed to cause change. Such studies might be diagramed as follows:

$$X \quad O$$

As has been pointed out (e.g., Boring, 1954; Stouffer, 1949) such studies have such a total absence of control as to be of almost no scientific value. The design is introduced here as a minimum reference point. Yet because of the continued investment in such studies and the drawing of causal inferences from them, some comment is required. Basic to scientific evidence (and to all knowledge-diagnostic processes including the retina of the eye) is the process of comparison, of recording differences, or of contrast. Any appearance of absolute knowledge, or intrinsic knowledge about singular isolated objects, is found to be illusory upon analysis. Securing scientific evidence involves making at least one comparison. For such a comparison to be useful, both sides of the comparison should be made with similar care and precision.

In the case studies of Design 1, a carefully studied single instance is implicitly compared with other events casually observed and remembered. The inferences are based upon general expectations of what the data would have been had the X not occurred, etc. Such studies often involve tedious collection of specific detail, careful observation, testing, and the like, and in such instances involve the error of *misplaced precision.* How much more valuable the study would be if the one set of observations were reduced by half and the saved effort directed to the study in equal detail of an appropriate comparison instance. It seems well-nigh unethical at the present time to allow, as theses or dissertations in education, case studies of this nature (i.e., involving a single group observed at one time only). "Standardized" tests in such case studies provide only very limited help, since the rival sources of difference other than X are so numerous as to render the "standard" reference group almost useless as a "control group." On the same grounds, the many uncontrolled sources of difference between a present case study and potential future ones which might be compared with it are so numerous as to make justification in terms of providing a bench mark for future studies also hopeless. In general, it would be better to apportion the descriptive effort between both sides of an interesting comparison.

Design 1, if taken in conjunction with the implicit "common-knowledge" comparisons, has most of the weaknesses of each of the subsequent designs. For this reason, the spelling out of these weaknesses will be left to those more specific settings.

2. The One-Group Pretest-Posttest Design

While this design is still widely used in educational research, and while it is judged as enough better than Design 1 to be worth doing where nothing better can be done (see the discussion of quasi-experimental designs below), it is introduced here as a "bad example" to illustrate several of the confounded extraneous variables that can jeopardize *internal* validity. These variables offer plausible hypotheses explaining an O_1—O_2 difference, rival to the hypothesis that X caused the difference:

$$O_1 \quad X \quad O_2$$

The first of these uncontrolled rival hypotheses is *history*. Between O_1 and O_2 many other change-producing events may have occurred in addition to the experimenter's X. If the pretest (O_1) and the posttest (O_2) are made on different days, then the events in between may have caused the difference. To become a *plausible* rival hypothesis, such an event should have occurred to most of the students in the group under study, say in some other class period or via a widely disseminated news story. In Collier's classroom study (conducted in 1940, but reported in 1944), while students were reading Nazi propaganda materials, France fell; the attitude changes obtained seemed more likely to be the result of this event than of the própaganda itself.[4] *History* becomes a more plausible rival explanation of change the longer the O_1—O_2 time lapse, and might be regarded as a trivial problem in an experiment completed within a one- or two-hour period, although even here, extraneous sources such as laughter, distracting events, etc., are to be looked for. Relevant to the variable *history* is the feature of *experimental isolation,* which can so nearly be achieved in many physical science laboratories as to render Design 2 acceptable for much of their research. Such effective experimental isolation can almost never be assumed in research on teaching methods. For these reasons a minus has been entered for Design 2 in Table 1 under *History*. We will classify with *history* a group of possible effects of season or of institutional-event schedule, although these might also be placed with *maturation*. Thus optimism might vary with seasons and anxiety with the semester examination schedule (e.g., Crook, 1937; Windle, 1954). Such effects might produce an O_1—O_2 change confusable with the effect of X.

A second rival variable, or class of variables, is designated *maturation*. This term is used here to cover all of those biological or

[4] Collier actually used a more adequate design than this, designated Design 10 in the present system.

TABLE 1

SOURCES OF INVALIDITY FOR DESIGNS 1 THROUGH 6

	Sources of Invalidity											
	Internal								External			
	History	Maturation	Testing	Instrumentation	Regression	Selection	Mortality	Interaction of Selection and Maturation, etc.	Interaction of Testing and X	Interaction of Selection and X	Reactive Arrangements	Multiple-X Interference
Pre-Experimental Designs:												
1. One-Shot Case Study *X O*	−	−				−	−			−		
2. One-Group Pretest-Posttest Design *O X O*	−	−	−	−	?	+	+	−	−	−	?	
3. Static-Group Comparison *X O* - - - - - - *O*	+	?	+	+	+	−	−	−		−		
True Experimental Designs:												
4. Pretest-Posttest Control Group Design *R O X O* *R O O*	+	+	+	+	+	+	+	+	−	?	?	
5. Solomon Four-Group Design *R O X O* *R O O* *R X O* *R O*	+	+	+	+	+	+	+	+	+	?	?	
6. Posttest-Only Control Group Design *R X O* *R O*	+	+	+	+	+	+	+	+	+	?	?	

Note: In the tables, a minus indicates a definite weakness, a plus indicates that the factor is controlled, a question mark indicates a possible source of concern, and a blank indicates that the factor is not relevant.

It is with extreme reluctance that these summary tables are presented because they are apt to be "too helpful," and to be depended upon in place of the more complex and qualified presentation in the text. No + or − indicator should be respected unless the reader comprehends why it is placed there. In particular, it is against the spirit of this presentation to create uncomprehended fears of, or confidence in, specific designs.

psychological processes which systematically vary with the passage of time, independent of specific external events. Thus between O_1 and O_2 the students may have grown older, hungrier, more tired, more bored, etc., and the obtained difference may reflect this process rather than *X*. In remedial education, which focuses on exceptionally disadvantaged persons, a process of "spontaneous remission," analogous to wound healing, may be mistaken for the specific effect of a remedial *X*. (Needless to say, such a remission is not regarded as "spontaneous" in any causal sense, but rather represents the cumulative effects of learning processes and environmental pressures of the total daily experience, which would be operating even if no X had been introduced.)

A third confounded rival explanation is the effect of *testing,* the effect of the pretest itself. On achievement and intelligence tests, students taking the test for a second time, or taking an alternate form of the test, etc., usually do better than those taking the test for the first time (e.g., Anastasi, 1958, pp. 190–191; Cane & Heim, 1950). These effects, as much as three to five IQ points on the average for naïve test-takers, occur without any instruction as to scores or items missed on the first test. For personality tests, a similar effect is noted, with second tests showing, in general, better adjustment, although occasionally a highly significant effect in the opposite direction is found (Windle, 1954). For attitudes toward minority groups a second test may show more prejudice, although the evidence is very slight (Rankin & Campbell, 1955). Obviously, conditions of anonymity, increased awareness of what answer is socially approved, etc., all would have a bearing on the direction of the result. For prejudice items under conditions of anonymity, the adaptation level created by the hostile statements presented may shift the student's expectations as to what kinds of attitudes are tolerable in the direction of greater hostility. In a signed personality or adjustment inventory, the initial administration partakes of a problem-solving situation in which the student attempts to discover the disguised purpose of the test. Having done this (or having talked with his friends about their answers to some of the bizarre items), he knows better how to present himself acceptably the second time.

With the introduction of the problem of test effects comes a distinction among potential measures as to their *reactivity*. This will be an important theme throughout this chapter, as will a general exhortation to use *nonreactive* measures wherever possible. It has long been a truism in the social sciences that the process of measuring may change that which is being measured. The test-retest gain would be one important aspect of such change. (Another, the interaction of testing and X, will be discussed with Design 4, below. Furthermore, these reactions to the pretest are important to avoid even where they have different effects for different examinees.) The reactive effect can be expected whenever the testing process is in itself a stimulus to change rather than a passive record of behavior. Thus in an experiment on therapy for weight control, the initial weigh-in might in itself be a stimulus to weight reduction, even without the therapeutic treatment. Similarly, placing observers in the classroom to observe the teacher's pretraining human relations skills may in itself change the teacher's mode of discipline. Placing a microphone on the desk may change the group interaction pattern, etc. In general, the more novel and motivating the test device, the more reactive one can expect it to be.

Instrumentation or "instrument decay" (Campbell, 1957) is the term used to indicate a fourth uncontrolled rival hypothesis. This term refers to autonomous changes in the measuring instrument which might account for an O_1—O_2 difference. These changes would be analogous to the stretching or fatiguing of spring scales, condensation in a cloud chamber, etc. Where human observers are used to provide O_1 and O_2, processes of learning, fatiguing, etc., within the observers will produce O_1—O_2 differences. If essays are being graded, the grading standards may shift between O_1 and O_2 (suggesting the control technique of shuffling the O_1 and O_2 essays together and having them graded without knowledge of which came first). If classroom participation is being observed, then the observers may be more skillful, or more blasé, on the second occasion. If parents are being interviewed, the interviewer's familiarity with the interview schedule and with the particular parents may produce shifts. A change in observers between O_1 and O_2 could cause a difference.

A fifth confounded variable in some instances of Design 2 is *statistical regression*. If, for example, in a remediation experiment, students are picked for a special experimental treatment because they do particularly poorly on an achievement test (which becomes for them the O_1), then on a subsequent testing using a parallel form or repeating the same test, O_2 for this group will almost surely average higher than did O_1. This dependable result is not due to any genuine effect of X, any test-retest practice effect, etc. It is rather a tautological aspect of the imperfect correlation between O_1 and O_2. Because errors of inference due to overlooking regression effects have been so troublesome in educational research, because the fundamental insight into their nature is so frequently missed even by students who have had advanced courses in modern statistics, and because in later discussions (e.g., of Design 10 and the ex post facto analysis) we will assume this knowledge, an elementary and old-fashioned exposition is undertaken here. Figure 1 presents some artificial data in which pretest and posttest for a whole population correlate .50, with no change in the group mean or variability. (The data were

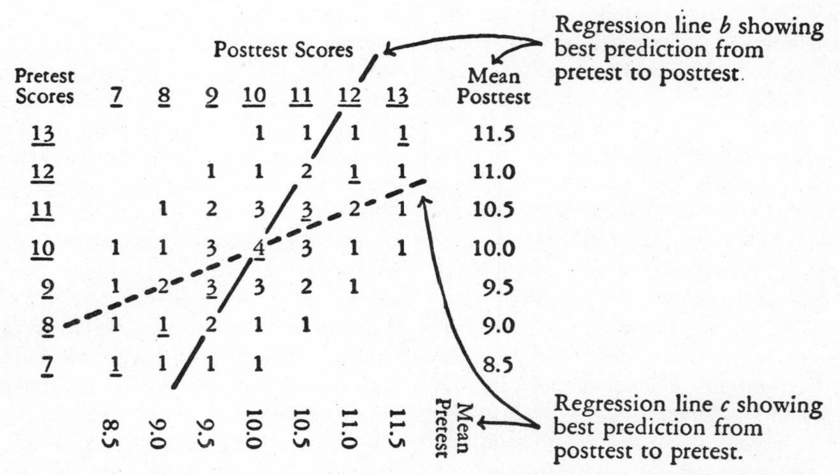

Pretest Scores	7	8	9	10	11	12	13	Mean Posttest
13				1	1	1	1	11.5
12			1	1	2	1	1	11.0
11		1	2	3	3	2	1	10.5
10	1	1	3	4	3	1	1	10.0
9	1	2	3	3	2	1		9.5
8	1	1	2	1	1			9.0
7	1	1	1	1				8.5
Mean Pretest	8.5	9.0	9.5	10.0	10.5	11.0	11.5	

Fig. 1*a*. Frequency Scatter of Posttest Scores for Each Class of Pretest Scores, and Vice Versa.

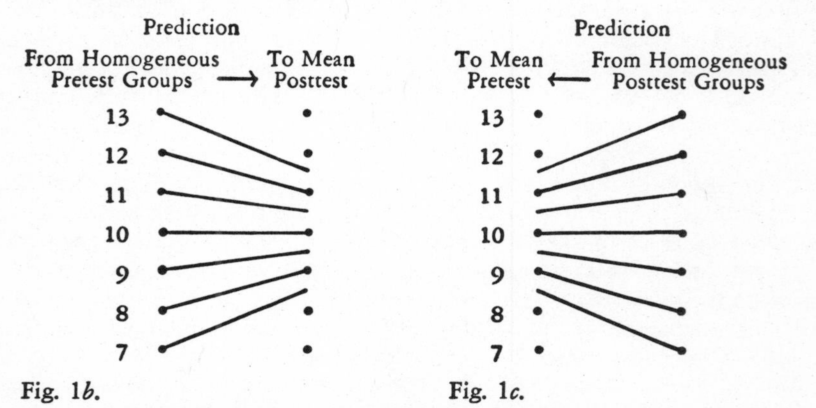

Fig. 1*b*.

Fig. 1*c*.

Fig. 1. Regression in the Prediction of Posttest Scores from Pretest, and Vice Versa.

selected to make the location of the row and column means obvious upon visual inspection. The value of .50 is similarly chosen for presentation convenience.) In this hypothetical instance, no true change has taken place, but as is usual, the fallible test scores show a retest correlation considerably less than unity. If, as suggested in the example initiated above, one starts by looking only at those with very low scores on the pretest, e.g., scores of 7, and looks only to the scores of these students on the posttest, one finds the posttest scores scattered, but in general better, and on the average "regressed" halfway (i.e., the regression or correlation coefficient is .50) back to the group mean, resulting in an average of 8.5. But instead of this being evidence of progress it is a tautological, if specific, restatement of the fact of imperfect correlation and its degree.

Because time passed and events occurred between pretest and posttest, one is tempted to relate this change causally to the specific direction of time passage. But note that a time-reversed analysis is possible here, as by starting with those whose posttest scores are 7, and looking at the scatter of their pretest scores, from which the reverse implication would be drawn—i.e., that scores are getting worse. The most mistaken causal inferences are drawn when the data are presented in the form of Fig. 1b (or the top or bottom portion of 1b). Here the bright appear to be getting duller, and the dull brighter, as if through the stultifying and homogenizing effect of an institutional environment. While this misinterpretation implies that the population variability on the posttest should be less than on the pretest, the two variabilities are in fact equal. Furthermore, by entering the analysis with pure groups of posttest scores (as in regression line c and Fig. 1c), we can draw the opposite inference. As McNemar (1940) pointed out, the use of time-reversed control analyses and the direct examination for changes in population variabilities are useful precautions against such misinterpretation.

We may look at regression toward the mean in another, related way. The more deviant the score, the larger the error of measurement it probably contains. Thus, in a sense, the typical extremely high scorer has had unusually good "luck" (large positive error) and the extremely low scorer bad luck (large negative error). Luck is capricious, however, so on a posttest we expect the high scorers to decline somewhat on the average, the low scorers to improve their relative standing. (The same logic holds if one begins with the posttest scores and works back to the pretest.)

Regression toward the mean is a ubiquitous phenomenon, not confined to pretesting and posttesting with the same test or comparable forms of a test. The principal who observes that his highest-IQ students tend to have less than the highest achievement-test score (though quite high) and that his lowest-IQ students are usually not right at the bottom of the achievement-test heap (though quite low) would be guilty of the regression fallacy if he declared that his school is understimulating the brightest pupils and overworking the dullest. Selecting those students who scored highest and lowest on the achievement test and looking at their IQs would force him by the same illogic to conclude the opposite.

While regression has been discussed here in terms of errors of measurement, it is more generally a function of the degree of correlation; the lower the correlation, the greater the regression toward the mean. The lack of perfect correlation may be due to "error" and/or to systematic sources of variance specific to one or the other measure.

Regression effects are thus inevitable accompaniments of imperfect test-retest correlation for groups *selected for their extremity*. They are not, however, necessary concomitants of extreme scores wherever encountered. If a group *selected for independent reasons* turns out to have an extreme mean, there is less a priori expectation that the group mean will regress on a second testing, for the random or extraneous sources of variance have been allowed to affect the initial scores in both directions. But for a group selected *because* of its extremity on a fallible variable, this is not the case. Its extremity is artificial and it will regress toward the mean of the population from which it was selected.

Regression effects of a more indirect sort can be due to selection of extreme scorers on measures other than the pretest. Consider a case in which students who "fail" a classroom examination are selected for experimental coaching. As a pretest, Form A of a standard achievement test is given, and as a posttest, Form B. It is probable that the classroom test correlates more highly with the immediate Form A administration than with the Form B administration some three months later (if the test had been given to the whole class on each occasion). The higher the correlation, the less regression toward the mean. Thus the classroom failures will have regressed upward less on the pretest than on the posttest, providing a pseudogain which might have been mistaken for a successful remedial-education effort. (For more details on gains and regression, see Lord, 1956, 1958; McNemar, 1958; Rulon, 1941; R. L. Thorndike, 1942.)

This concludes the list of weaknesses of Design 2 which can be conveniently discussed at this stage. Consulting Table 1 shows that there is one more minus under internal validity, for a factor which will not be examined until the discussion of Design 10 (see page 217) in the quasi-experimental designs section, and two minuses for external validity, which will not be explained until the discussion of Design 4 (see page 186).

3. The Static-Group Comparison

The third pre-experimental design needed for our development of invalidating factors is the static-group comparison. This is a design in which a group which has experienced X is compared with one which has not, for the purpose of establishing the effect of X.

$$\frac{X \;\; - \; - \;\; O_1}{\qquad\qquad O_2}$$

Instances of this kind of research include, for example, the comparison of school systems which require the bachelor's degree of teachers (the X) versus those which do not; the comparison of students in classes given speed-reading training versus those not given it; the comparison of those who heard a certain TV program with those who did not, etc. In marked contrast with the "true" experiment of Design 6, below, there are in these Design 3 instances no formal means of certifying that the groups would have been equivalent had it not been for the X. This absence, indicated in the diagram by the dashed lines separating the two groups, provides the next factor needing control, i.e., *selection*. If O_1 and O_2 differ, this difference could well have come about through the differential recruitment of persons making up the groups: the groups might have differed anyway, without the occurrence of X. As will be discussed below under the ex post facto analysis, matching on background characteristics other than O is usually ineffective and misleading, particularly in those instances in which the persons in the "experimental group" have sought out exposure to the X.

A final confounded variable for the present list can be called experimental *mortality*, or the production of O_1—O_2 differences in groups due to the differential drop-out of persons from the groups. Thus, even if in Design 3 the two groups had once been identical, they might differ now not because of any change on the part of individual members, but rather because of the selective drop-out of persons from one of the groups. In educational research this problem is most frequently met in those studies aimed at ascertaining the effects of a college education by comparing measures on freshmen (who have not had the X) with seniors (who have). When such studies show freshman women to be more beautiful than senior women, we recoil from the implication that our harsh course of training is debeautifying, and instead point to the hazards in the way of a beautiful girl's finishing college before getting married. Such an effect is classified here as experimental *mortality*. (Of course, if we consider the *same* girls when they are freshmen and seniors, this problem disappears, and we have Design 2.)

THREE TRUE EXPERIMENTAL DESIGNS

The three basic designs to be treated in this section are the currently recommended designs in the methodological literature. They will also turn out to be the most strongly recommended designs of this presentation, even though this endorsement is subject to many specific qualifications regarding usual practice and to some minus signs in Table 1 under *external validity*. Design 4 is the most used of the three, and for this reason we allow its presentation to be disproportionately extended and to become the locus of discussions more generally applicable. Note that all three of these designs are presented in terms of a single X being compared with *no* X. Designs with more numerous treatments in the Fisher factorial experiment tradition represent important elaborations tangential to the main thread of this chapter and are discussed at the end of this section, subsequent to Design 6. But this perspective can serve to remind us at this point that the comparison of X with *no* X is an oversimplification. The comparison is actually with the specific activities of the control group which have filled the time period corresponding to that in which the experimental group receives the X. Thus the comparison might better be between X_1 and X_c, or between X_1 and X_o, or X_1 and X_2. That these control group activities are often unspecified adds an undesirable ambiguity to the interpretation of the contribution of X. Bearing these comments in mind, we will continue in this section the graphic convention of presenting no X in the control group.

4. The Pretest-Posttest Control Group Design

Controls for Internal Validity

One or another of the above considerations led psychological and educational researchers between 1900 and 1920 to add a control group to Design 2, creating the presently orthodox control group design. McCall (1923), Solomon (1949), and Boring (1954) have given us some of this history, and a scanning of the *Teachers College Record* for that period implies still more, for as early as 1912 control groups were being referred to without need of explanation (e.g., Pearson, 1912). The control group designs thus introduced are classified in this chapter under two heads: the present Design 4 in which equivalent groups as achieved by randomization are employed, and the quasi-experimental Design 10 in which extant intact comparison groups of unassured equivalence are employed. Design 4 takes this form:

$$\begin{array}{llll} R & O_1 & X & O_2 \\ R & O_3 & & O_4 \end{array}$$

Because the design so neatly controls for *all* of the seven rival hypotheses described so far, the presentations of it have usually not made explicit the control needs which it met. In the tradition of learning research, the practice effects of *testing* seem to provide the first recognition of the need for a control group. *Maturation* was a frequent critical focus in experimental studies in education, as well as in the nature-nurture problem in the child development area. In research on attitude change, as in the early studies on the effects of motion pictures, *history* may have been the main necessitating consideration. In any event, it seems desirable here to discuss briefly the way in which, or the conditions under which, these factors are controlled.

History is controlled insofar as general historical events that might have produced an O_1—O_2 difference would also produce an O_3—O_4 difference. Note, however, that many supposed utilizations of Design 4 (or 5 or 6) do *not* control for unique *intrasession history*. If all of the randomly assigned students in the experimental group are treated in a single session, and similarly the control students in another single session, then the irrelevant unique events in either session (the obstreperous joke, the fire across the street, the experimenter's introductory remarks, etc.) become rival hypotheses explaining the O_1—O_2 versus O_3—O_4 difference. Such an experiment is *not* a true experiment, even when presented, as was Solomon's (1949) experiment on the teaching of spelling, as an illustrative paradigm. (To be fair, we point out that it was chosen to illustrate a different point.) Thinking over our "best practice" on this point may make this seem a venial sin, but our "best practice" is producing experiments too frequently unreplicable, and this very source of "significant" but extraneous differences might well be an important fault. Furthermore, the typical experiment in the *Journal of Experimental Psychology* does achieve control of intrasession history through testing students and animals individually and through assigning the students and experimental periods at random to experimental or control conditions. Note, however, that even with individual sessions, history can be uncontrolled if all of the experimental group is run before the control group, etc. Design 4 calls for simultaneity of experimental and control sessions. If we actually run sessions simultaneously, then different experimenters must be used, and experimenter differences can become a form of intrasession history confounded with X.

The optimal solution is a randomization of experimental occasions, with such restrictions as are required to achieve balanced representation of such highly likely sources of bias as experimenters, time of day, day of week, portion of semester, nearness to examinations, etc. The common expedient of running experimental subjects in small groups rather than individually is inadmissible if this grouping is disregarded in the statistical analysis. (See the section on assigning intact groups to treatments, below.) All those in the same session share the same intrasession history, and thus have sources of similarity other than X. If such sessions have been assigned at random, the correct statistical treatment is the same as that discussed below for the assignment of intact classrooms to treatments. (For some studies involving group testing, the several experimental treatments can be randomly distributed within one face-to-face group, as in using multiple test forms in a study of the effect of the order of difficulty of items. In such cases, the specificities of intrasession history are common to both treatments and do not become a plausible rival hypothesis confounded with X in explaining the differences obtained.)

Maturation and *testing* are controlled in that they should be manifested equally in experimental and control groups. *Instrumentation* is easily controlled where the conditions for the control of intrasession history are met, particularly where the O is achieved by student responses to a fixed instrument such as a printed test. Where observers or interviewers are used, however, the problem becomes more serious. If observers are few enough not to be randomly assignable to the observation of single sessions, then not only should each observer be used for both experimental and control sessions, but in addition, the observers should be kept ignorant as to which students are receiving which treatments, lest the knowledge bias their ratings or records. That such bias tendencies are "dependable" sources of variance is affirmed by the necessity in medical research of the second blind in the double-blind experiment, by recent research (Rosenthal, 1959), and by older studies (e.g., Kennedy & Uphoff, 1939; Stanton & Baker, 1942). The use of recordings of group interaction, so that judges may judge a series of randomized sections of pretest, posttest, experimental, and control group transcriptions, helps to control instrumentation in research on classroom behavior and group interaction.

Regression is controlled as far as mean differences are concerned, no matter how extreme the group is on pretest scores, if both experimental and control groups are randomly assigned from this same extreme pool. In such a case, the control group regresses as much as does the experimental group. Interpretative lapses due to regression artifacts do frequently occur, however, even under Design 4 conditions. An experimenter may employ the control group to confirm group mean effects of *X*, and then abandon it while examining which pretest-score subgroups of the experimental group were most influenced. If the whole group has shown a gain, then he arrives at the stimulating artifact that those initially lowest have gained most, those initially highest perhaps not at all. This outcome is assured because under conditions of total group mean gain, the regression artifact supplements the gain score for the below-mean pretest scorers, and tends to cancel it for the high pretest scorers. (If there was no over-all gain, then the experimenter may mistakenly "discover" that this was due to two mutually cancelling effects, for those low to gain, those high to lose.) One cure for these misinterpretations is to make parallel analyses of extreme pretest scorers in the control group, and to base differential gain interpretations on comparisons of the posttest scores of the corresponding experimental and control pretest subgroups. (Note, however, that skewed distributions resulting from selection make normal-curve statistics of dubious appropriateness.)

Selection is ruled out as an explanation of the difference to the extent that randomization has assured group equality at time *R*. This extent is the extent stated by our sampling statistics. Thus the assurance of equality is greater for large numbers of random assignments than for small. To the extent indicated by the error term for the no-difference hypothesis, this assumption will be wrong occasionally. In Design 4, this means that there will occasionally be an apparently "significant" difference between the pretest scores. Thus, while simple or stratified randomization assures unbiased assignment of experimental subjects to groups, it is a less than perfect way of assuring the initial equivalence of such groups. It is nonetheless the only way of doing so, and the essential way. This statement is made so dogmatically because of a widespread and mistaken preference in educational research over the past 30 years for equation through matching. McCall (1923) and Peters and Van Voorhis (1940) have helped perpetuate this misunderstanding. As will be spelled out in more detail in the discussion of Design 10 and the ex post facto analysis below, matching is no real help when used to overcome initial group differences. This is not to rule out matching as an adjunct to randomization, as when one gains statistical precision by assigning students to matched pairs, and then randomly assigning one member of each pair to the experimental group, the other to the control group. In the statistical literature this is known as "blocking." See particularly the discussions of Cox (1957), Feldt (1958), and Lindquist (1953). But matching as a substitute for randomization is taboo even for the quasi-experimental designs using but two natural intact groups, one experimental, the other control: even in this weak "experiment," there are better ways than matching for attempting to correct for initial mean differences in the two samples.

The data made available by Design 4 make it possible to tell whether *mortality* offers a plausible explanation of the O_1—O_2 gain. Mortality, lost cases, and cases on which only partial data are available, are troublesome to handle, and are commonly swept under the rug. Typically, experiments on teaching methods are spread out over days, weeks, or months. If the pretests and posttests are given in the classrooms from which experimental group and control group are drawn, and if the experimental condition requires attendance at certain sessions, while the control condition does not, then the differential attendance on the three occasions (pretest, treatment, and posttest) produces "mortality" which can introduce subtle sample biases.

If, of those initially designated as experimental group participants, one eliminates those who fail to show up for experimental sessions, then one selectively shrinks the experimental group in a way not comparably done in the control group, biasing the experimental group in the direction of the conscientious and healthy. The preferred mode of treatment, while not usually employed, would seem to be to use all of the selected experimental and control students who completed both pretest and posttest, including those in the experimental group who failed to get the *X*. This procedure obviously attenuates the apparent effect of the *X*, but it avoids the sampling bias. This procedure rests on the assumption that no simpler mortality biases were present; this assumption can be partially checked by examining both the number and the pretest scores of those who were present on pretest but not on posttest. It is possible that some *X*s would affect this drop-out rate rather than change individual scores. Of course, even where drop-out rates are the same, there remains the possibility of complex interactions which would tend to make the character of the drop-outs in the experimental and control groups differ.

The mortality problem can be seen in a greatly exaggerated form in the *invited remedial treatment* study. Here, for example, one sample of poor readers in a high school is invited to participate in voluntary remedial sessions, while an equivalent group are not invited. Of the invited group, perhaps 30 per cent participate. Posttest scores, like pretest scores, come from standard reading achievement tests administered to all in the classrooms. It is unfair to compare the 30 per cent volunteers with the total of the control group, because they represent those most disturbed by their pretest scores, those likely to be most vigorous in self-improvement, etc. But it is impossible to locate their exact counterparts in the control group. While it also seems unfair to the hypothesis of therapeutic effectiveness to compare the total invited group with the total uninvited group, this is an acceptable, if conservative, solution. Note, however, the possibility that the invitation itself, rather than the therapy, causes the effect. In general, the uninvited control group should be made just as aware of its standing on the pretest as is the invited group. Another alternative is to invite all those who need remedial sessions and to assign those who accept into true and placebo remedial treatment groups; but in the present state of the art, any placebo therapy which is plausible enough to look like help to the student is apt to be as good a therapy as is the treatment we are studying. Note, however, the valid implication that experimental tests of the relative efficacy of two therapeutic procedures are much easier to evaluate than the absolute effectiveness of either. The only solution in actual use is that of creating experimental and control groups from among seekers of remedial treatment by manipulating waiting periods (e.g., Rogers & Dymond, 1954). This of course sometimes creates other difficulties, such as an excessive drop-out from the postponed-therapy control group. For a successful and apparently nonreactive use of a lottery to decide on an immediate or next-term remedial reading course, see Reed (1956).

Factors Jeopardizing External Validity

The factors of internal invalidity which have been described so far have been factors which directly affected *O* scores. They have been factors which by themselves could produce changes which might be mistaken for the results of *X*, i.e., factors which, once the control group was added, would produce effects manifested by themselves in the control group and added onto the effects of *X* in the experimental group. In the language of analysis of variance, *history, maturation, testing,* etc., have been described as main effects, and as such have been controlled in Design 4, giving it *internal* validity. The threats to *external* validity, on the other hand, can be called interaction effects, involving *X* and some other variable. They thus represent a potential specificity of the effects of X to some undesirably limited set of conditions. To anticipate: in Design 4, for all we know, the effects of X observed may be specific to groups warmed up by the pretest. We are logically unable to generalize to the larger unpretested universe about which we would prefer to be able to speak.

In this section we shall discuss several such threats to generalizability, and procedures for reducing them. Thus since there are valid designs avoiding the pretest, and since in many settings (but not necessarily in research on teaching) it is to unpretested groups that one wants to generalize, such designs are preferred on grounds of *external* validity or generalizability. In the area of teaching, the doubts frequently expressed as to the applicability in actual practice of the results of highly artificial experiments are judgments about *external* validity. The introduction of such considerations into the discussion of optimal experimental designs thus strikes a sympathetic note in the practitioner who rightly feels that these considerations have been unduly neglected in the usual formal treatise on experimental methodology. The ensuing discussion will support such views by pointing out numerous ways in which experiments can be made more valid externally, more appropriate bases of generalization to teaching practice, without losing *internal* validity.

But before entering this discussion, a caveat is in order. This caveat introduces some painful problems in the science of induction. The problems are painful because of a recurrent reluctance to accept Hume's truism that *induction or generalization is never fully justified logically*. Whereas the problems of *internal* validity are solvable within the limits of the logic of probability statistics, the problems of external validity are not logically solvable in any neat, conclusive way. Generalization always turns out to involve extrapolation into a realm not represented in one's sample. Such extrapolation is made by *assuming* one knows the relevant laws. Thus, if one has an internally valid Design 4, one has demonstrated the effect only for those specific conditions which the experimental and control group have in common, i.e., only for pretested groups of a specific age, intelligence, socioeconomic status, geographical region, historical moment, orientation of the stars, orientation in the magnetic field, barometric pressure, gamma radiation level, etc.

Logically, we cannot generalize beyond these limits; i.e., we cannot generalize at all. But we do attempt generalization by guessing at laws and checking out some of these generalizations in other equally specific but different conditions. In the course of the history of a science we learn about the "justification" of generalizing by the cumulation of our experience in generalizing, but this is not a logical generalization deducible from the details of the original experiment. Faced by this, we do, in generalizing, make guesses as to yet unproven laws, including some not even explored. Thus, for research on teaching, we are quite willing to assume that orientation in the magnetic field has no effect. But we know from scattered research that pretesting has often had an effect, and therefore we would like to remove it as a limit to our generalization. If we were doing research on iron bars, we would know from experience that an initial weighing has never been found to be reactive, but that orientation in magnetic field, if not systematically controlled, might seriously limit the generalizability of our discoveries. The sources of external invalidity are thus guesses as to general laws in the science of a science: guesses as to what factors lawfully interact with our treatment variables, and, by implication, guesses as to what can be disregarded.

In addition to the specifics, there is a general empirical law which we are assuming, along with all scientists. This is the modern version of Mill's assumption as to the lawfulness of nature. In its modern, weaker version, this can be stated as the assumption of the "stickiness" of nature: we assume that the closer two events are in time, space, and measured value on any or all dimensions, the more they tend to follow the same laws.

While complex interactions and curvilinear relationships are expected to confuse attempts at generalization, they are more to be expected the more the experimental situation differs from the setting to which one wants to generalize. Our call for greater external validity will thus be a call for that maximum similarity of experiments to the conditions of application which is compatible with internal validity.

While stressing this, we should keep in mind that the "successful" sciences such as physics and chemistry made their strides without any attention to representativeness (but with great concern for repeatability by independent researchers). An ivory-tower artificial laboratory science is a valuable achievement even if unrepresentative, and artificiality may often be essential to the analytic separation of variables fundamental to the achievements of many sciences. But certainly, if it does not interfere with internal validity or analysis, external validity is a very important consideration, especially for an applied discipline such as teaching.

Interaction of testing and X. In discussions of experimental design per se, the threat of the pretest to external validity was first presented by Solomon (1949), although the same considerations had earlier led individual experimenters to the use of Design 6, which omits the pretest. Especially in attitude-change studies, where the attitude tests themselves introduce considerable amounts of unusual content (e.g., one rarely sees in cold print as concentrated a dose of hostile statements as is found in the typical prejudice test), it is quite likely that the person's attitudes and his susceptibility to persuasion are changed by a pretest. As a psychologist, one seriously doubts the comparability of one movie audience seeing *Gentlemen's Agreement* (an antiprejudice film) immediately after having taken a 100-item anti-Semitism test with another audience seeing the movie without such a pretest. These doubts extend not only to the main effect of the pretest, but also to its effect upon the response to persuasion. Let us assume that that particular movie was so smoothly done that some persons could enjoy it for its love interest without becoming aware of the social problem it dealt with. Such persons would probably not occur in a pretested group. If a pretest sensitized the audience to the problem, it might, through a focusing of attention, increase the educational effect of the *X*. Conceivably, such an *X* might be effective only for a pretested group.

While such a sensitizing effect is frequently mentioned in anecdotal presentations of the effect, the few published research results show either no effect (e.g., Anderson, 1959; Duncan, et al., 1957; Glock, 1958; Lana, 1959a, 1959b; Lana & King, 1960; Piers, 1955; Sobol, 1959; Zeisel, 1947) or an interaction effect of a dampening order. Thus Solomon (1949) found that giving a pretest reduced the efficacy of experimental spelling training, and Hovland, Lumsdaine, and Sheffield (1949) suggested that a pretest reduced the persuasive effects of movies. This interaction effect is well worth avoiding, even if not as misleading as sensitization (since false positives are more of a problem in our literature than false negatives, owing to the glut of published findings [Campbell, 1959, pp. 168–170]).

The effect of the pretest upon *X* as it restricts external validity is of course a function of the extent to which such repeated measurements are characteristic of the universe to which one wants to generalize. In the area of mass communications, the researcher's interview and attitude-test procedures are quite atypical. But in research on teaching, one is interested in generalizing to a setting in which testing is a regular phenomenon. Especially if the experiment can use regular classroom examinations as *O*s, but probably also if the experimental *O*s are similar to those usually used, no undesirable interaction of *testing* and *X* would be present. Where highly unusual test procedures are used, or where the testing procedure involves deception, perceptual or cognitive restructuring, surprise, stress, etc., designs having unpretested groups remain highly desirable if not essential.

Interaction of selection and X. While Design 4 controls for the effects of selection at the level of explaining away experimental and control group differences, there remains the possibility that the effects validly demonstrated hold only for that unique population from which the experimental and control groups were jointly selected. This possibility becomes more likely as we have more difficulty in getting subjects for our experiment. Consider the implications of an experiment on teaching in which the researcher has been turned down by nine school systems and is finally accepted by a tenth. This tenth almost certainly differs from the other nine, and from the universe of schools to which we would like to generalize, in many specific ways. It is, thus, nonrepresentative. Almost certainly its staff has higher morale, less fear of being inspected, more zeal for improvement than does that of the average school. And the effects we find, while internally valid, might be specific to such schools. To help us judge on these matters, it would seem well for research reports to include statements as to how many and what kind of schools and classes were asked to cooperate but refused, so that the reader can estimate the severity of possible selective biases. Generally speaking, the greater the amount of cooperation involved, the greater the amount of disruption of routine, and the higher our refusal rate, the more opportunity there is for a selection-specificity effect.

Let us specify more closely just what the "*interaction* of selection and *X*" means. If we were to conduct a study within a single volunteered school, using random assignment of subjects to experimental and control groups, we would not be concerned about the "main effect" of the school itself. If both experimental and control group means were merely elevated equally by this, then no harm would be done. If, however, there were characteristics of the school that caused the experimental treatment to be more effective there than it would be in the target population of schools, this could be serious. We want to know that the interaction of school characteristics (probably related to voluntarism) with experimental treatments is negligible. Some experimental variables might be quite sensitive to (interact with) school characteristics; others might not. Such interaction *could* occur between schools with similar mean IQs, or it could be absent when IQ differences were great. We would expect, however, that interactions would be more likely if the schools differed markedly in various characteristics than if they were similar.

Often stringent sampling biases occur because of the inertia of experimenters who do not allow a more representative selection of schools the opportunity to refuse to participate. Thus most research on teaching is done in those schools with the highest percentage of university professors' children enrolled. While sampling representativeness is impossible of perfect achievement and is almost totally neglected in many sciences (in most studies appearing in the *Journal of Experimental Psychology,* for example), it both can and should be emphasized as a desideratum in research on teaching. One way to increase it is to reduce the number of students or classrooms participating from a given school or grade and to increase the number of schools and grades in which the experiment is carried on. It is obvious that we are never going to conduct experiments on samples representatively drawn from all United States classrooms, or all world classrooms. We will learn how far we can generalize an internally valid finding only piece by piece through trial and error of generalization efforts. But these generalization efforts will succeed more often if in the initial experiment we have demonstrated the phenomenon over a wide variety of conditions.

With reference to the pluses and minuses of Table 1, it is obvious that nothing firm can be entered in this column. The column is presented, however, because the requirements of some designs exaggerate or ameliorate this problem. Design 4 in the social-attitudes realm is so demanding of cooperation on the part of respondents or subjects as to end up with research done only on captive audiences rather than the general citizen of whom one would wish to speak. For such a setting, Design 4 would rate a minus for selection. Yet for research on teaching, our universe of interest is a captive population, and for this, highly representative Design 4s can be done.

Other interactions with X. In parallel fashion, the interaction of *X* with the other factors can be examined as threats to *external* validity. Differential *mortality* would be a product of *X* rather than interactive with it. *Instrumentation* interacting with *X* has been implicitly included in the discussion of *internal* validity, since an instrumentation effect specific to the presence of *X* would counterfeit a true effect of *X* (e.g., where observers make ratings, know the hypothesis, and know which students have received *X*). A threat to external validity is the possibility of the specificity of effects to the specific instruments (tests, observers, meters, etc.) used in the study. If multiple observers or interviewers are used across treatments, such interactions can be studied directly (Stanley, 1961a). *Regression* does not enter as interacting with *X*.

Maturation has implications of a selection-specificity nature: the results may be specific to those of this given age level, fatigue level, etc. The interaction of *history* and *X* would imply that the effect was specific to the historical conditions of the experiment, and while validly observed there, would not be found upon other occasions. The fact that the experiment was done during wartime, or just following an unsuccessful teachers' strike, etc., might produce a responsiveness to *X* not to be found upon other occasions. If we were to produce a sampling model for this problem, we should want the experiment replicated over a random sample of past and future occasions, which is obviously impossible. Furthermore, we share with other sciences the empirical assumption that there are no truly time-dependent laws, that the effects of *history* where found will be due to the specific combinations of stimulus conditions at that time, and thus ultimately will be incorporated under time-independent general laws (Neyman, 1960). ("Expanding universe" cosmologies may seem to require qualification of this statement, but not in ways relevant to this discussion.) Nonetheless, successful replication of research results across times as well as settings increases our confidence in a generalization by making interaction with *history* less likely.

These several factors have not been entered as column headings in Table 1, because they do not provide bases of discrimination among alternative designs.

Reactive arrangements. In the usual psychological experiment, if not in educational research, a most prominent source of unrepresentativeness is the patent artificiality of the experimental setting and the student's knowledge that he is participating in an experiment. For human experimental subjects, a higher-order problem-solving task is generated, in which the procedures and experimental treatment are reacted to not only for their simple stimulus values, but also for their role as clues in divining the experimenter's intent. The play-acting, outguessing, up-for-inspection, I'm-a-guinea-pig, or whatever attitudes so generated are unrepresentative of the school setting, and seem to be qualifiers of the effect of *X*, seriously hampering generalization. Where such *reactive arrangements* are unavoidable, internally valid experiments of this type should by all means be continued. But if they can be avoided, they obviously should be. In stating this, we in part join the typical anti-experimental critic in the school system or the education faculty by endorsing his most frequent protest as to the futility of "all this research." Our more moderate conclusion is not, however, that research should be abandoned for this reason, but rather that it should be improved on this score. Several suggestions follow.

Any aspect of the experimental procedure may produce this *reactive arrangements* effect. The pretesting in itself, apart from its contents, may do so, and part of the *pretest* interaction with *X* may be of this nature, although there are ample grounds to suspect the content features of the testing process. The process of randomization and assignment to treatments may be of such a nature: consider the effect upon a classroom when (as in Solomon, 1949) a randomly selected half of the pupils in a class are sent to a separate room. This action, plus the presence of the strange "teachers," must certainly create expectations of the unusual, with wonder and active puzzling as to purpose. The presentation of the treatment *X*, if an out-of-ordinary event, could have a similar effect. Presumably, even the posttest in a posttest-only Design 6 could create such attitudes. The more obvious the connection between the experimental treatment and the posttest content, the more likely this effect becomes.

In the area of public opinion change, such reactive arrangements may be very hard to avoid. But in much research on teaching methods there is no need for the students to know that an experiment is going on. (It would be nice to keep the teachers from knowing this, too, in analogy to medicine's double-blind experiment, but this is usually not feasible.) Several features may make such disguise possible. If the *X*s are variants on usual classroom events occurring at plausible periods in the curriculum calendar, then one-third of the battle is won when these treatments occur without special announcement. If the *O*s are similarly embedded as regular examinations, the second requirement is achieved. If the *X*s are communications focused upon individual students, then randomization can be achieved without the physical transportation of randomly equivalent samples to different classrooms, etc.

As a result of such considerations, and as a result of personal observations of experimenters who have published data in spite of having such poor rapport that their findings were quite misleading, the present authors are gradually coming to the view that experimentation within schools must be conducted by regular staff of the schools concerned, whenever possible, especially when findings are to be generalized to other classroom situations.

At present, there seem to be two main types of "experimentation" going on within schools: (1) research "imposed" upon the school by an outsider, who has his own ax to grind and whose goal is not immediate action (change) by the school; and (2) the so-called "action" researcher, who tries to get teachers themselves to be "experimenters," using that word quite loosely. The first researcher gets results that may be rigorous but not applicable. The latter gets results that may be highly applicable but probably not "true" because of extreme lack of rigor in the research. An alternative model is for the ideas for classroom research to originate with teachers and other school personnel, with designs to test these ideas worked out cooperatively with specialists in research methodology, and then for the bulk of the experimentation to be carried out by the idea-producers themselves. The appropriate statistical analyses could be done by the research methodologist and the results fed back to the group via a trained intermediary (supervisor, director of research in the school system, etc.) who has served as intermediary all along. Results should then be relevant and "correct." How to get *basic* research going under such a pattern is largely an unsolved problem, but studies could become less and less ad hoc and more and more theory-oriented under a competent intermediary.

While there is no intent in this chapter to survey either good or bad examples in the literature, a recent study by Page (1958) shows such an excellent utilization of these features (avoiding reactive arrangements, achieving sampling representativeness, and avoiding testing-*X* interactions) that it is cited here as a concrete illustration of optimal practice. His study shows that brief written comments upon returned objective examinations improve subsequent objective examination performance. This finding was demonstrated across 74 teachers, 12 school systems, 6 grades (7–12), 5 performance levels (A, B, C, D, F), and a wide variety of subjects, with almost no evidence of interaction effects. The teachers and classes were randomly selected. The earliest regular objective examination in each class was used as the pretest. By rolling a specially marked die the teacher assigned students to treatment groups, and correspondingly put written comments on the paper or did not. The next normally scheduled objective test in the class became the posttest. As far as could be told, not one of the 2,139 students was aware of experimentation. Few instructional procedures lend themselves to this inconspicuous randomization, since usually the oral communication involved is addressed to a whole class, rather than to individuals. (Written communications do allow for randomized treatment, although student detection of varied treatments is a problem.) Yet, holding these ideals in mind, research workers can make experiments nonreactive in many more features than they are at present.

Through regular classroom examinations or through tests presented as regular examinations and similar in content, and through alternative teaching procedures presented without announcement or apology in the regular teaching process, these two sources of reactive arrangements can probably be avoided in most instances. Inconspicuous randomization may be the more chronic problem. Sometimes, in large high schools or colleges, where students sign up for popular courses at given hours and are then assigned arbitrarily to multiple simultaneous sections, randomly equivalent sections might be achieved through control of the assignment process. (See Siegel & Siegel, 1957, for an opportunistic use of a natural randomization process.) However, because of unique intragroup histories, such initially equivalent sections become increasingly nonequivalent with the passage of long periods of time.

The all-purpose solution to this problem is to move the randomization to the classroom as a unit, and to construct experimental and control groups each constituted of numerous classrooms randomly assigned (see Lindquist, 1940, 1953). Usually, but not essentially, the classrooms would be classified for analysis on the basis of such factors as school, teacher (where teachers have several classes), subject, time of day, mean intelligence level, etc.; from these, various experimental-treatment groups would be assigned by a random process. There have been a few such studies, but soon they ought to become standard. Note that the appropriate test of significance is *not* the pooling of all students as though the students had been assigned at random. The details will be discussed in the subsequent section.

Tests of Significance for Design 4

Good experimental design is separable from the use of statistical tests of significance. It is the art of achieving interpretable comparisons and as such would be required even if the end product were to be graphed percentages, parallel prose case studies, photographs of groups in action, etc. In all such cases, the interpretability of the "results" depends upon control over the factors we have been describing. If the comparison is interpretable, then statistical tests of significance come in for the decision as to whether or not the obtained difference rises above the fluctuations to be expected in cases of no true difference for samples of that size. Use of significance tests presumes but does not prove or supply the comparability of the comparison groups or the interpretability of the difference found. We would thus be happy to teach experimental design upon the grounds of common sense and nonmathematical considerations. We hope that the bulk of this chapter is accessible to students of education still lacking in statistical training. Nevertheless, the issue of statistical procedures is intimately tied to experimental design, and we therefore offer these segregated comments on the topic. (Also see Green & Tukey, 1960; Kaiser, 1960; Nunnally, 1960; and Rozeboom, 1960.)

A wrong statistic in common use. Even though Design 4 is the standard and most widely used design, the tests of significance used with it are often wrong, incomplete, or inappropriate. In applying the common "critical ratio" or *t* test to this standard experimental design, many researchers have computed two *t*s, one for the pretest-posttest difference in the experimental group, one for the pretest-posttest gain in the control group. If the former be "statistically significant" and the latter "not," then they have concluded that the *X* had an effect, without any direct statistical comparison of the experimental and control groups. Often the conditions have been such that, had a more appropriate test been made, the difference would not have been significant (as in the case where the significance values are borderline, with the control group showing a gain almost reaching significance). Windle (1954) and Cantor (1956) have shown how frequent this error is.

Use of gain scores and covariance. The most widely used acceptable test is to compute for each group pretest-posttest gain scores and to compute a *t* between experimental and control groups on these gain scores. Randomized "blocking" or "leveling" on pretest scores and the analysis of covariance with pretest scores as the covariate are usually preferable to simple gain-score comparisons. Since the great bulk of educational experiments show no significant difference, and hence are frequently not reported, the use of this more precise analysis would seem highly desirable. Considering the labor of conducting an experiment, the labor of doing the proper analysis is relatively trivial. Standard treatments of Fisher-type analyses may be consulted for details. (Also see Cox, 1957, 1958; Feldt, 1958; and Lindquist, 1953.)

Statistics for random assignment of intact classrooms to treatments. The usual statistics are appropriate only where individual students have been assigned at random to treatments. Where intact classes have been assigned to treatments, the above formulas would provide too small an error term because the randomization procedure obviously has been more "lumpy" and fewer chance events have been employed. Lindquist (1953, pp. 172–189) has provided the rationale and formulas for a correct analysis. Essentially, the class means are used as the basic observations, and treatment effects are tested against variations in these means. A covariance analysis would use pretest means as the covariate.

Statistics for internal validity. The above points were introduced to convey the statistical orthodoxy relevant to experimental design. The point to follow represents an effort to expand or correct that orthodoxy. It extends an implication of the distinction between *external* and *internal validity* over into the realm of sampling statistics. The statistics discussed above all imply sampling from an infinitely large universe, a sampling more appropriate to a public opinion survey than to the usual laboratory experiment. In the rare case of a study like Page's (1958), there is an actual sampling from a large predesignated universe, which makes the usual formulas appropriate. At the other extreme is the laboratory experiment represented in the *Journal of Experimental Psychology,* for example, in which *internal validity* has been the only consideration, and in which *all* members of a unique small universe have been exhaustively assigned to the treatment groups. There is in such experiments a great emphasis upon randomization, but not for the purpose of securing representativeness for some larger population. Instead, the randomization is solely for the purpose of equating experimental and control groups or the several treatment groups. The randomization is thus within a very small finite population which is in fact the sum of the experimental plus control groups.

This extreme position on the sampling universe is justified when describing laboratory procedures of this type: volunteers are called for, with or without promises of rewards in terms of money, personality scores, course credit points, or completion of an obligatory requirement which they will have to meet sometime during the term anyway. As volunteers come in, they are randomly assigned to treatments. When some fixed number of subjects has been reached, the experiment is stopped. There has not even been a random selection from within a much larger list of volunteers. Early volunteers are a biased sample, and the total universe "sampled" changes from day to day as the experiment goes on, as more pressure is required to recruit volunteers, etc. At some point the procedure is stopped, all designatable members of the universe having been used in one or another treatment group. Note that the sampling biases implied do not in the least jeopardize the random equivalence of the treatment groups, but rather only their "representativeness."

Or consider a more conscientious scientist, who randomly draws 100 names from his lecture class of 250 persons, contacting them by phone or mail, and then as they meet appointments assigns them randomly to treatment groups. Of course, some 20 of them cannot conveniently be fitted into the laboratory time schedule, or are ill, etc., so a redefinition of the universe has taken place implicitly. And even if he doggedly gets all 100, from the point of view of representativeness, what he has gained is the ability to generalize with statistical confidence to the 1961 class of Educational Psychology A at State Teachers. This new universe, while larger, is not intrinsically of scientific interest. Its bounds are not the bounds specified by any scientific theory. The important interests in generalization will have to be explored by the sampling of other experiments elsewhere. Of course, since his students are less select, there is more external validity, but not enough gain to be judged worth it by the great bulk of experimental psychologists.

In general, it is obvious that the dominant purpose of randomization in laboratory experiments is internal validity, not external. Pursuant to this, more appropriate and smaller error terms based upon small finite universes should be employed. Following Kempthorne (1955) and Wilk and Kempthorne (1956), we note that the appropriate model is urn randomization, rather than sampling from a universe. Thus there is available a more appropriate, more precise, nonparametric test, in which one takes the obtained experimental and control group scores and repeatedly assigns them at random to two "urns," generating empirically (or mathematically) a distribution of mean differences arising wholly from random assignment of these particular scores. This distribution is the criterion with which the obtained mean difference should be compared. When "plot-treatment interaction" (heterogeneity of true effects among subjects) is present, this distribution will have less variability than the corresponding distribution assumed in the usual t test.

These comments are not expected to modify greatly the actual practice of applying tests of significance in research on teaching. The exact solutions are very tedious, and usually inaccessible. Urn randomization, for example, ordinarily requires access to high-speed computers. The direction of error is known: using the traditional statistics is too conservative, too inclined to say "no effect shown." If we judge our publications to be overloaded with "false-positives," i.e., claims for effects that won't hold up upon cross-validation (this is certainly the case for experimental and social psychology, if not as yet for research on teaching), this error is in the preferred direction—if error there must be. Possible underestimation of significance is greatest when there are only two experimental conditions and all available subjects are used (Wilk & Kempthorne, 1955, p. 1154).

5. The Solomon Four-Group Design

While Design 4 is more used, Design 5, the Solomon (1949) Four-Group Design, deservedly has higher prestige and represents the first explicit consideration of *external validity* factors. The design is as follows:

$$\begin{array}{llll} R & O_1 & X & O_2 \\ R & O_3 & & O_4 \\ R & & X & O_5 \\ R & & & O_6 \end{array}$$

By paralleling the Design 4 elements (O_1 through O_4) with experimental and control groups lacking the pretest, both the main effects of *testing* and the interaction of *testing* and X are determinable. In this way, not only is generalizability increased, but in addition, the effect of X is replicated in four different fashions: $O_2 > O_1$, $O_2 > O_4$, $O_5 > O_6$, and $O_5 > O_3$. The actual instabilities of experimentation are such that if these comparisons are in agreement, the strength of the inference is greatly increased. Another indirect contribution to the generalizability of experimental findings is also made, in that through experience with Design 5 in any given research area one learns the general likelihood of testing-by-X interactions, and thus is better able to interpret past and future Design 4s. In a similar way, one can note (by comparison of O_6 with O_1 and O_3) a combined effect of maturation and history.

Statistical Tests for Design 5

There is no singular statistical procedure which makes use of all six sets of observations simultaneously. The asymmetries of the design rule out the analysis of variance of gain scores. (Solomon's suggestions concerning these are judged unacceptable.) Disregarding the pretests, except as another "treatment" coordinate with X, one can treat the posttest scores with a simple 2×2 analysis of variance design:

	No X	X
Pretested	O_4	O_2
Unpretested	O_6	O_5

From the column means, one estimates the main effect of X, from row means, the main effect of pretesting, and from cell means, the interaction of testing with X. If the main and interactive effects of pretesting are negligible, it may be desirable to perform an analysis of covariance of O_4 versus O_2, pretest scores being the covariate.

6. The Posttest-Only Control Group Design

While the pretest is a concept deeply embedded in the thinking of research workers in education and psychology, it is not actually essential to true experimental designs. For psychological reasons it is difficult to give up "knowing for sure" that the experimental and control groups were "equal" before the differential experimental treatment. Nonetheless, the most adequate all-purpose assurance of lack of initial biases between groups is randomization. Within the limits of confidence stated by the tests of significance, randomization can suffice without the pretest. Actually, almost all of the agricultural experiments in the Fisher (1925, 1935) tradition are without pretest. Furthermore, in educational research, particularly in the primary grades, we must frequently experiment with methods for the initial introduction of entirely new subject matter, for which pretests in the ordinary sense are impossible, just as pretests on believed guilt or innocence would be inappropriate in a study of the effects of lawyers' briefs upon a jury. Design 6 fills this need, and in addition is appropriate to all of the settings in which Designs 4 or 5 might be used, i.e., designs where true randomization is possible. Its form is as follows:

$$\begin{array}{lll} R & X & O_1 \\ R & & O_2 \end{array}$$

While this design was used as long ago as the 1920's, it has not been recommended in most methodological texts in education. This has been due in part to a confusion of it with Design 3, and due in part to distrust of randomization as equation. The design can be considered as the two last groups of the Solomon Four-Group Design, and it can be seen that it controls for testing as main effect and interaction, but unlike Design 5 it does not measure them. However, such measurement is tangential to the central question of whether or not X did have an effect. Thus, while Design 5 is to be preferred to Design 6 for reasons given above, the extra gains from Design 5 may not be worth the more than double effort. Similarly, Design 6 is usually to be preferred to Design 4, unless there is some question as to the genuine randomness of the assignment. Design 6 is greatly underused in educational and psychological research.

However, in the repeated-testing setting of much educational research, if appropriate antecedent variates are available, they should certainly be used for blocking or leveling, or as covariates. This recommendation is made for two reasons: first, the statistical tests available for Design 4 are more powerful than those available for Design 6. While the greater effort of Design 4 outweighs this gain for most research settings, it would not do so where suitable antecedent scores were automatically available. Second, the availability of pretest scores makes possible examination of the interaction of *X* and pretest ability level, thus exploring the generalizability of the finding more thoroughly. Something similar can be done for Design 6, using other available measures in lieu of pretests, but these considerations, coupled with the fact that for educational research frequent testing is characteristic of the universe to which one wants to generalize, may reverse the case for generally preferring Design 6 over Design 4. Note also that for any substantial mortality between *R* and the posttest, the pretest data of Design 4 offer more opportunity to rule out the hypothesis of differential mortality between experimental and control groups.

Even so, many problems exist for which pretests are unavailable, inconvenient, or likely to be reactive, and for such purposes the legitimacy of Design 6 still needs emphasis in many quarters. In addition to studies of the mode of teaching novel subject materials, a large class of instances remains in which (1) the *X* and posttest *O* can be delivered to students or groups as a single natural package, and (2) a pretest would be awkward. Such settings frequently occur in research on testing procedures themselves, as in studies of different instructions, different answer-sheet formats, etc. Studies of persuasive appeals for volunteering, etc., are similar. Where student anonymity must be kept, Design 6 is usually the most convenient. In such cases, randomization is handled in the mixed ordering of materials for distribution.

The Statistics for Design 6

The simplest form would be the *t* test. Design 6 is perhaps the only setting for which this test is optimal. However, covariance analysis and blocking on "subject variables" (Underwood, 1957b) such as prior grades, test scores, parental occupation, etc., can be used, thus providing an increase in the power of the significance test very similar to that provided by a pretest. Identicalness of pretest and posttest is not essential. Often these will be different forms of "the same" test and thus less identical than a repetition of the pretest. The gain in precision obtained corresponds directly to the degree of covariance, and while this is usually higher for alternate forms of "the same" test than for "different" tests, it is a matter of degree, and something as reliable and factorially complex as a grade-point average might turn out to be superior to a short "pretest." Note that a grade-point average is not usually desirable as a posttest measure, however, because of its probable insensitivity to *X* compared with a measure more specifically appropriate in content and timing. Whether such a pseudo pretest design should be classified as Design 6 or Design 4 is of little moment. It would have the advantages of Design 6 in avoiding an experimenter-introduced pretest session, and in avoiding the "giveaway" repetition of identical or highly similar unusual content (as in attitude change studies). It is for such reasons that the entry for Design 6 under "reactive arrangements" should be slightly more positive than that for Designs 4 and 5. The case for this differential is, of course, much stronger for the social sciences in general than for research on educational instruction.

Factorial Designs

On the conceptual base of the three preceding designs, but particularly of Designs 4 and 6, the complex elaborations typical of the Fisher factorial designs can be extended by adding other groups with other Xs. In a typical single-classification criterion or "one-way" analysis of variance we would have several "levels" of the treatment, e.g., X_1, X_2, X_3, etc., with perhaps still an X_0 (no-X) group. If the control group be regarded as one of the treatments, then for Designs 4 and 6 there would be one group for each treatment. For Design 5 there would be two groups (one pretested, one not) for each treatment, and a two-classification ("two-way") analysis of variance could still be performed. We are not aware that more-than-two-level designs of the Design 5 type have been done. Usually, if one were concerned about the pretest interaction, Design 6 would be employed because of the large number of groups otherwise required. Very frequently, two or more treatment variables, each at several "levels," will be employed, giving a series of groups that could be designated X_{a1} X_{b1}, X_{a1} X_{b2}, X_{a1} X_{b3}, . . ., X_{a2} X_{b1}, etc.

Such elaborations, complicated by efforts to economize through eliminating some of the possible permutations of X_a by X_b, have produced some of the traumatizing mysteries of factorial design (randomized blocks, split plots, Greco-Latin squares, fractional replication, confounding, etc.) which have created such a gulf between advanced and traditional research methodologies in education. We hope that this chapter helps bridge this gulf through continuity with traditional methodology and the common-sense considerations which the student brings with him. It is also felt that a great deal of what needs to be taught about experimental design can best be understood when presented in the form of two-treatment designs, without interference from other complexities. Yet a full presentation of the problems of traditional usage will generate a comprehension of the need for and place of the modern approaches. Already, in searching for the most efficient way of summarizing the widely accepted old-fashioned Design 4, we were introduced to a need for covariance analysis, which has been almost unused in this setting. And in Design 5, with a two-treatment problem elaborated only to obtain needed controls, we moved away from critical ratios or t tests into the related analysis-of-variance statistics.

The details of statistical analyses for factorial designs cannot be taught or even illustrated in this chapter. Elementary aspects of these methods are presented for educational researchers by Edwards (1960), Ferguson (1959), Johnson and Jackson (1959), and Lindquist (1953). It is hoped, however, that the ensuing paragraphs may convey some understanding of certain alternatives and complexities particularly relevant for the design issues discussed in this chapter. The complexities to be discussed do not include the common reasons for using Latin squares and many other incomplete designs where knowledge concerning certain interactions is sacrificed merely for reasons of cost. (But the use of Latin squares as a substitute for control groups where randomization is not possible will be discussed as quasi-experimental Design 11 below.) The reason for the decision to omit such incomplete designs is that detailed knowledge of interactions is highly relevant to the external validity problem, particularly in a science which has experienced trouble in replicating one researcher's findings in another setting (see Wilk & Kempthorne, 1957). The concepts which we seek to convey in this section are interaction, nested versus crossed classifications, and finite, fixed, random, and mixed factorial models.

Interaction

We have already used this concept in contexts where it was hoped the untrained

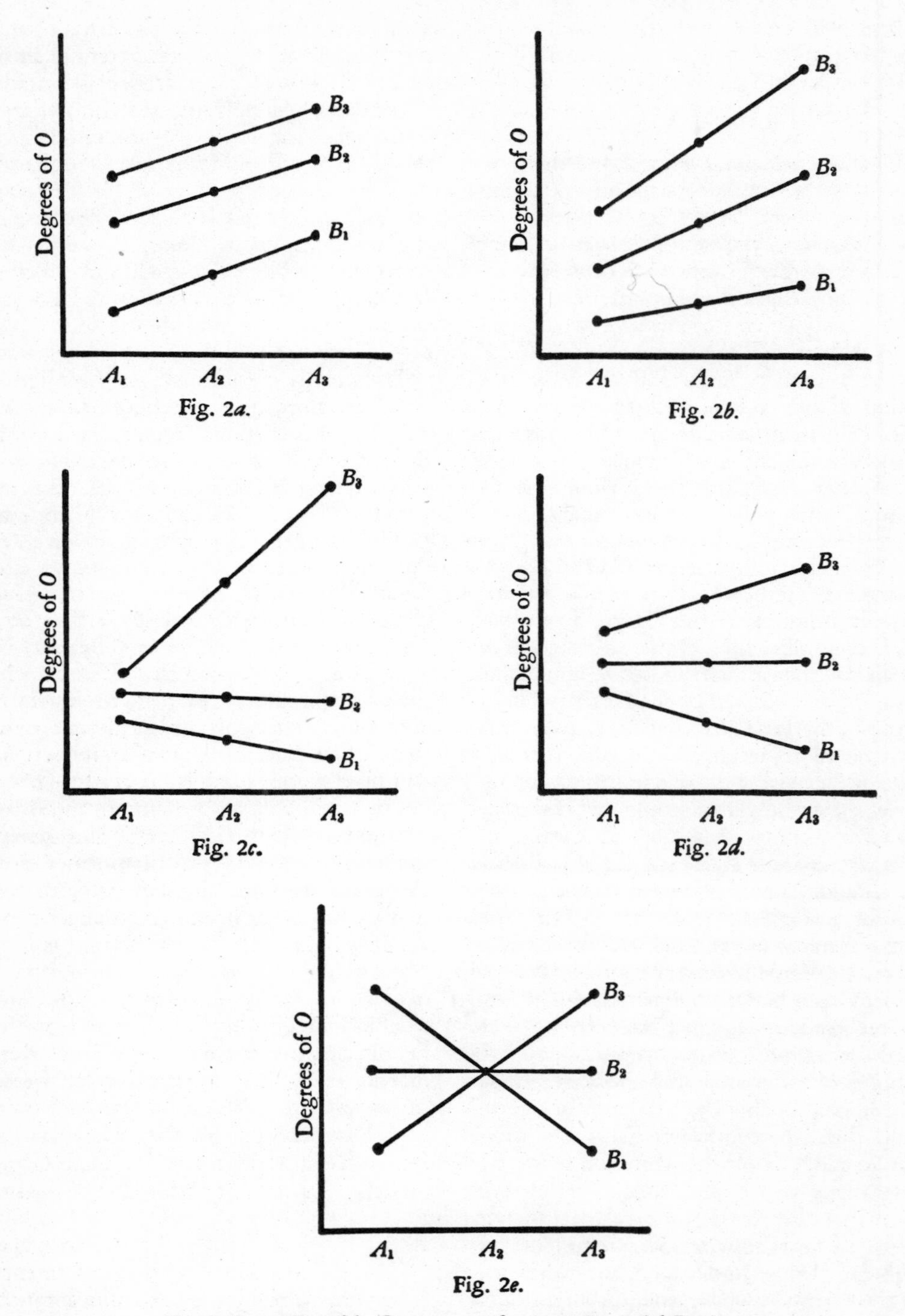

Fig. 2. Some Possible Outcomes of a 3 x 3 Factorial Design.

reader would find it comprehensible. As before, our emphasis here is upon the implications for generalizability. Let us consider in graphic form, in Fig. 2, five possible outcomes of a design having three levels each of X_a and X_b, to be called here *A* and *B*. (Since three dimensions [*A*, *B*, and *O*] are to be graphed in two dimensions, there are several alternative presentations, only one of which is used here.) In Fig. 2*a* there is a significant main effect for both *A* and *B*, but no interaction. (There is, of course, a summation of effects—A_3, B_3 being strongest—but no interaction, as the effects are additive.) In all of the others, there are significant interactions in addition to, or instead of, the main effects of *A* and *B*. That is, the law as to the effect of *A* changes depending upon the specific value of *B*. In this sense, interaction effects are specificity-of-effect rules and are thus relevant to generalization efforts. The interaction effect in 2*d* is most clearly of this order. Here *A* does not have a main effect (i.e., if one averages the values of all three *B*s for each *A*, a horizontal line results). But when *B* is held at level 1, increases in *A* have a decremental effect, whereas when *B* is held at level 3, *A* has an incremental effect. Note that had the experimenter varied *A* only and held *B* constant at level 1, the results, while internally valid, would have led to erroneous generalizations for B_2 and B_3. The multiple-factorial feature of the design has thus led to valuable explorations of the generalizability or external validity of any summary statement about the main effect of *A*. Limitations upon generalizability, or specificity of effects, appear in the statistical analysis as significant interactions.

Figure 2*e* represents a still more extreme form of interaction, in which neither *A* nor *B* has any main effect (no general rules emerge as to which level of either is better) but in which the interactions are strong and definite. Consider a hypothetical outcome of this sort. Let us suppose that three types of teachers are all, in general, equally effective (e.g., the spontaneous extemporizers, the conscientious preparers, and the close supervisors of student work). Similarly, three teaching methods in general turn out to be equally effective (e.g., group discussion, formal lecture, and tutorial). In such a case, even in the absence of "main effects" for either teacher-type or teaching method, teaching methods could plausibly interact strongly with types, the spontaneous extemporizer doing best with group discussion and poorest with tutorial, and the close supervisor doing best with tutorial and poorest with group discussion methods.

From this point of view, we should want to distinguish between the kinds of significant interactions found. Perhaps some such concept as "monotonic interactions" might do. Note that in 2*b*, as in 2*a*, there is a main effect of both *A* and *B*, and that *A* has the same directional effect in every separate panel of *B* values. Thus we feel much more confident in generalizing the expectation of increase in *O* with increments in *A* to novel settings than we do in case 2*c*, which likewise might have significant main effects for *A* and *B*, and likewise a significant *A*-*B* interaction. We might, in fact, be nearly as confident of the generality of *A*'s main effect in a case like 2*b* as in the interaction-free 2*a*. Certainly, in interpreting effects for generalization purposes, we should plot them and examine them in detail. Some "monotonic" or single-directional interactions produce little or no specificity limitations. (See Lubin, 1961, for an extended discussion of this problem.)

Nested Classifications

In the illustrations which we have given up to this point, all of the classification criteria (the *A*s and the *B*s) have "crossed" all other classification criteria. That is, all levels of *A* have occurred with all levels of *B*. Analysis of variance is not limited to this situation, however.

So far, we have used, as illustrations, classification criteria which were "experimental treatments." Other types of classification criteria, such as sex and age of pupils, could be introduced into many experiments as fully crossed classifications. But to introduce the most usual uses of "nested" classifications, we must present the possibility of less obvious classification criteria. One of these is "teachers." Operating at the fully crossed level, one might do an experiment in a high school in which each of 10 teachers used each of two methods of teaching a given subject, to different experimental classes. In this case, teachers would be a fully crossed classification criterion, each teacher being a different "level." The "main effect" of "teachers" would be evidence that some teachers are better than others no matter which method they are using. (Students or classes must have been assigned at random; otherwise teacher idiosyncrasies and selection differences are confounded.) A significant interaction between teachers and methods would mean that the method which worked better depended upon the particular teacher being considered.

Suppose now, in following up such an interaction, one were interested in whether or not a given technique was, in general, better for men teachers than women. If we now divide our 10 teachers into 5 men and 5 women, a "nesting" classification occurs in that the teacher classification, while still useful, does not cross sexes; i.e., the same teacher does not appear in both sexes, while each teacher and each sex do cross methods. This nesting requires a somewhat different analysis than does the case where all classifications cross all others. (For illustrative analyses, see Green and Tukey, 1960, and Stanley, 1961a.) In addition, certain interactions of the nested variables are ruled out. Thus the teachers-sex and teachers-sex-method interactions are not computable, and, indeed, make no sense conceptually.

"Teachers" might also become a nested classification if the above experiment were extended into several schools, so that schools became a classification criterion (for which the main effects might reflect learning-rate differences on the part of pupils of the several schools). In such a case, teachers would usually be "nested" within schools, in that one teacher would usually teach classes within just one school. While in this instance a teacher-school interaction is conceivable, one could not be computed unless all teachers taught in both schools, in which case teachers and schools would be "crossed" rather than "nested."

Pupils, or subjects in an experiment, can also be treated as a classification criterion. In a fully crossed usage each pupil gets each treatment, but in many cases the pupil enters into several treatments, but not all; i.e., nesting occurs. One frequent instance is the study of trial-by-trial data in learning. In this case, one might have learning curves for each pupil, with pupils split between two methods of learning. Pupils would cross trials but not methods. Trial-method interactions and pupil-trial interactions could be studied, but not pupil-method interactions. Similarly, if pupils are classified by sex, nesting occurs.

Most variables of interest in educational experimentation can cross other variables and need not be nested. Notable exceptions, in addition to those mentioned above, are chronological age, mental age, school grade (first, second, etc.), and socioeconomic level. The perceptive reader may have noted that independent variables, or classification criteria, are of several sorts: (1) manipulated variables, such as teaching method, assignable at will by the experimenter; (2) potentially manipulable aspects, such as school subject studied, that the experimenter might assign in some random way to the pupils he is using, but rarely does; (3) relatively fixed aspects of the environment, such as community or school or socioeconomic level, not under the direct control of the experimenter but serving as explicit bases for stratification in the experiment; (4) "organismic" characteristics of pupils, such as age, height, weight, and sex; and (5) response characteristics of pupils, such as scores on various tests. Usually the manipulated independent variables of Class 1 are of primary interest, while the unmanipulated independent variables of Classes 3, 4, and sometimes 5, serve to increase precision and reveal how generalizable the effects of manipulated variables are. The variables of Class 5 usually appear as covariates or dependent variates. Another way to look at independent variables is to consider them as intrinsically ordered (school grade, socioeconomic level, height, trials, etc.) or unordered (teaching method, school subject, teacher, sex, etc.). Effects of ordered variables may often be analyzed further to see whether the trend is linear, quadratic, cubic, or higher (Grant, 1956; Myers, 1959).

Finite, Random, Fixed, and Mixed Models

Recently, stimulated by Tukey's unpublished manuscript of 1949, several mathematical statisticians have devised "finite" models for the analysis of variance that apply to the sampling of "levels" of experimental factors (independent variables) the principles well worked out previously for sampling from finite populations. Scheffé (1956) provided a historical survey of this clarifying development. Expected mean squares, which help determine appropriate "error terms," are available (Stanley, 1956) for the completely randomized three-classification factorial design. Finite models are particularly useful because they may be generalized readily to situations where one or more of the factors are random or fixed. A simple explanation of these extensions was given by Ferguson (1959).

Rather than present formulas, we shall use a verbal illustration to show how finite, random, and fixed selection of levels of a factor differ. Suppose that "teachers" constitute one of several bases for classification (i.e., independent variables) in an experiment. If 50 teachers are available, we might draw 5 of these *randomly* and use them in the study. Then a factor-sampling coefficient $(1-5/50)$, or 0.9, would appear in some of our formulas. If all 50 teachers were employed, then teachers would be a "fixed" effect and the coefficient would become $(1-50/50) = 0$. If, on the other hand, a virtually infinite population of teachers existed, 50 selected randomly from this population would be an infinitesimal percentage, so the coefficient would approach 1 for each "random" effect. The above coefficients modify the formulas for expected mean squares, and hence for "error" terms. Further details appear in Brownlee (1960), Cornfield and Tukey (1956), Ferguson (1959), Wilk and Kempthorne (1956), and Winer (1962).

Other Dimensions of Extension

Before leaving the "true" experiments for the quasi-experimental designs, we wish to explore some other extensions from this simple core, extensions appropriate to all of the designs to be discussed.

Testing for Effects Extended in Time

In the area of persuasion, an area somewhat akin to that of educating and teaching, Hovland and his associates have repeatedly found that long-term effects are not only quantitatively different, but also qualitatively different. Long-range effects are greater than immediate effects for general attitudes, although weaker for specific attitudes (Hovland, Lumsdaine, & Sheffield, 1949). A discredited speaker has no persuasive effect immediately, but may have a significant effect a month later, unless listeners are reminded of the source (Hovland, Janis, & Kelley, 1953). Such findings warn us against pinning all of our experimental evaluation of teaching methods on immediate posttests or measures at any single point in time. In spite of the immensely greater problems of execution (and the inconvenience to the nine-month schedule for a Ph.D. dissertation), we can but recommend that posttest periods such as one month, six months, and one year be included in research planning.

When the posttest measures are grades and examination scores that are going to be collected anyway, such a study is nothing but a bookkeeping (and mortality) problem. But where the *O*s are introduced by the experimenter, most writers feel that repeated posttest measures on the same students would be more misleading than the pretest would be. This has certainly been found to be true in research on memory (e.g., Underwood, 1957a). While Hovland's group has typically used a pretest (Design 4), they have set up separate experimental and control groups for each time delay for the posttest, e.g.:

R	*O*	*X*	*O*	
R	*O*		*O*	
R	*O*	*X*		*O*
R	*O*			*O*

A similar duplication of groups would be required for Designs 5 or 6. Note that this design lacks perfect control for its purpose of comparing differences in effect as a function of elapsed time, in that the differences could also be due to an interaction between *X* and the specific historical events occurring between the short-term posttest and the long-term one. Full control of this possibility leads to still more elaborate designs. In view of the great expense of such studies except where the *O*s are secured routinely, it would seem incumbent upon those making studies using institutionalized *O*s repeatedly available to make use of the special advantages of their settings by following up the effects over many points in time.

Generalizing to Other *X*s: Variability in the Execution of *X*

The goal of science includes not only generalization to other populations and times but also to other nonidentical representations of the treatment, i.e., other representations which theoretically should be the same, but which are not identical in theoretically irrelevant specifics. This goal is contrary to an often felt extension of the demand for experimental control which leads to the desire for an *exact* replication of the *X* on each repetition. Thus, in studying the effect of an emotional versus a rational appeal, one might have the same speaker give all appeals to each type of group or, more extremely, record the talks so that all audiences of a given treatment heard "exactly the same" message. This might seem better than having several persons give each appeal just once, since in the latter case we "would not know exactly" what experimental stimulus each session got. But the reverse is actually the case, if by "know" we mean the ability to pick the proper abstract classification for the treatment and to convey the information effectively to new users. With the taped interview we have repeated each time many specific irrelevant features; for all we know, these details, not the intended features, created the effect. If, however, we have many independent exemplifications, the specific irrelevancies are not apt to be repeated each time, and our interpretation of the source of the effects is thus more apt to be correct.

For example, consider the Guetzkow, Kelly, and McKeachie (1954) comparison of recitation and discussion methods in teaching. Our "knowledge" of what the experimental treatments were, in the sense of being able to draw recommendations for other teachers, is better *because* eight teachers were used, each interpreting each method in his own way, than if only one teacher had been used, or than if the eight had memorized common details not included in the abstract description of the procedures under comparison. (This emphasis upon heterogeneous execution of *X* should if possible be accompanied, as in Guetzkow, et al., 1954, by having each treatment executed by each of the experimental teachers, so that no specific irrelevancies are confounded with a specific treatment. To estimate the significance of teacher-method interaction when intact classes have been employed, each teacher should execute each method twice.)

In a more obvious illustration, a study of the effect of sex of the teacher upon beginning instruction in arithmetic should use numerous examples of each sex, not just one of each. While this is an obvious precaution, it has not always been followed, as Hammond (1954) has pointed out. The problem is an aspect of Brunswik's (1956) emphasis upon representative design. Underwood (1957b, pp. 281–287) has on similar grounds argued against the exact standardization or the exact replication of apparatus from one study to another, in a fashion not incompatible with his vigorous operationalism.

Generalizing to Other *X*s: Sequential Refinement of *X* and Novel Control Groups

The actual *X* in any experiment is a complex package of what will eventually be conceptualized as several variables. Once a strong and clear-cut effect has been noted, the course of science consists of further experiments which refine the *X,* teasing out those aspects which are most essential to the effect. This refinement can occur through more specifically defined and represented treatments, or through developing novel control groups, which come to match the experimental group on more and more features of the treatment, reducing the differences to more specific features of the original complex *X*. The placebo control group and the sham-operation control group in medical research illustrate this. The prior experiments demonstrated an internally valid effect, which, however, could have been due to the patient's knowledge that he was being treated or to surgical shock, rather than to the specific details of the drug or to the removal of the brain tissue—hence the introduction of the special controls against these possibilities. The process of generalizing to other *X*s is an exploratory, theory-guided trial and error of extrapolations, in the process of which such refinement of *X*s is apt to play an important part.

Generalizing to Other *O*s

Just as a given *X* carries with it a baggage of theoretically irrelevant specificities which may turn out to cause the effect, so any given *O,* any given measuring instrument, is a complex in which the relevant content is necessarily embedded in a specific instrumental setting, the details of which are tangential to the theoretical purpose. Thus, when we use IBM pencils and machine-scored answer-sheets, it is usually for reasons of convenience and not because we wish to include in our scores variance due to clerical skills, test-form familiarity, ability to follow instructions, etc. Likewise, our examination of specific subject-matter competence by way of essay tests must be made through the vehicles of penmanship and vocabulary usage and hence must contain variance due to these sources often irrelevant to our purposes. Given this inherent complexity of any *O,* we are faced with a problem when we wish to generalize to other potential *O*s. To which aspect of our experimental *O* was this internally valid effect due? Since the goals of teaching are not solely those of preparing people for future essay and objective examinations, this problem of external validity or generalizability is one which must be continually borne in mind.

Again, conceptually, the solution is not to hope piously for "pure" measures with no irrelevant complexities, but rather to use multiple measures in which the specific vehicles, the specific irrelevant details, are as different as possible, while the common content of our concern is present in each. For *O*s, more of this can be done within a single experiment than for *X*s, for it is usually possible to get many measures of effect (i.e., dependent variables) in one experiment. In the study by Guetzkow, Kelly, and McKeachie (1954), effects were noted not only on course examinations and on special attitude tests introduced for this purpose, but also on such subsequent behaviors as choice of major and enrollment in advanced courses in the same topic. (These behaviors proved to be just as sensitive to treatment differences as were the test measures.) *Multiple Os should be an orthodox requirement in any study of teaching methods.* At the simplest level, both essay and objective examinations should be used (see Stanley & Beeman, 1956), along with indices of classroom participation, etc., where feasible. (An extension of this perspective to the question of test validity is provided by Campbell and Fiske, 1959; and Campbell, 1960.)

QUASI-EXPERIMENTAL DESIGNS[5]

There are many natural social settings in which the research person can introduce something like experimental design into his scheduling of data collection procedures (e.g., the *when* and *to whom* of measurement), even though he lacks the full control over the scheduling of experimental stimuli (the *when* and *to whom* of exposure and the ability to randomize exposures) which makes a true experiment possible. Collectively, such situations can be regarded as quasi-experimental designs. One purpose of this chapter is to encourage the utilization of such quasi-experiments and to increase awareness of the kinds of settings in which opportunities to employ them occur. But just because full experimental control *is* lacking, it becomes imperative that the researcher be thoroughly aware of which specific variables his particular design fails to control. It is for this need in evaluating quasi-experiments, more than for understanding true experiments, that the check lists of sources of invalidity in Tables 1, 2, and 3 were developed.

The average student or potential researcher reading the previous section of this chapter probably ends up with more things to worry about in designing an experiment than he had in mind to begin with. This is all to the good if it leads to the design and execution of better experiments and to more circumspection in drawing inferences from results. It is, however, an unwanted side effect if it creates a feeling of hopelessness with regard to achieving experimental control and leads to the abandonment of such efforts in favor of even more informal methods of investigation. Further, this formidable list of sources of invalidity might, with even more likelihood, reduce willingness to undertake quasi-experimental designs, designs in which from the very outset it can be seen that full experimental control is lacking. Such an effect would be the opposite of what is intended.

From the standpoint of the final interpretation of an experiment and the attempt to fit it into the developing science, every experiment is imperfect. What a check list of validity criteria can do is to make an experimenter more aware of the residual imperfections in his design so that on the relevant points he can be aware of competing interpretations of his data. He should, of course, design the very best experiment which the situation makes possible. He should deliberately seek out those artificial and natural laboratories which provide the best opportunities for control. But beyond that he should go ahead with experiment and interpretation, fully aware of the points on which the results are equivocal. While this awareness is important for experiments in which "full" control has been exercised, it is crucial for quasi-experimental designs.

In implementing this general goal, we shall in this portion of the chapter survey the strengths and weaknesses of a heterogeneous collection of quasi-experimental designs, each deemed worthy of use *where better designs are not feasible*. First will be discussed three single-group experimental designs. Following these, five general types of multiple-group experiments will be presented. A separate section will deal with correlation, ex post facto designs, panel studies, and the like.

SOME PRELIMINARY COMMENTS ON THE THEORY OF EXPERIMENTATION

This section is written primarily for the educator who wishes to take his research

[5] This section draws heavily upon D. T. Campbell, Quasi-experimental designs for use in natural social settings, in D. T. Campbell, *Experimenting, Validating, Knowing: Problems of Method in the Social Sciences.* New York: McGraw-Hill, in preparation.

out of the laboratory and into the operating situation. Yet the authors cannot help being aware that experimental psychologists may look with considerable suspicion on any effort to sanction studies having less than full experimental control. In part to justify the present activity to such monitors, the following general comments on the role of experiments in science are offered. These comments are believed to be compatible with most modern philosophies of science, and they come from a perspective on a potential general psychology of inductive processes (Campbell, 1959).

Science, like other knowledge processes, involves the proposing of theories, hypotheses, models, etc., and the acceptance or rejection of these on the basis of some external criteria. Experimentation belongs to this second phase, to the pruning, rejecting, editing phase. We may assume an ecology for our science in which the number of potential positive hypotheses very greatly exceeds the number of hypotheses that will in the long run prove to be compatible with our observations. *The task of theory-testing data collection is therefore predominantly one of rejecting inadequate hypotheses.* In executing this task, any arrangement of observations for which certain outcomes would disconfirm theory will be useful, including quasi-experimental designs of less efficiency than true experiments.

But, it may be asked, will not such imperfect designs result in spurious confirmation of inadequate theory, mislead our subsequent efforts, and waste our journal space with the dozens of studies which it seems to take to eradicate one conspicuously published false positive? This is a serious risk, but a risk which we must take. It is a risk shared in kind, if not in the same degree, by "true" experiments of Designs 4, 5, and 6. In a very fundamental sense, experimental results never "confirm" or "prove" a theory—rather, the successful theory is tested and escapes being disconfirmed. The word "prove," by being frequently employed to designate deductive validity, has acquired in our generation a connotation inappropriate both to its older uses and to its application to inductive procedures such as experimentation. The results of an experiment "probe" but do not "prove" a theory. An adequate hypothesis is one that has repeatedly survived such probing—but it may always be displaced by a new probe.

It is by now generally understood that the "null hypothesis" often employed for convenience in stating the hypothesis of an experiment can never be "accepted" by the data obtained; it can only be "rejected," or "fail to be rejected." Similarly with hypotheses more generally—they are technically never "confirmed": where we for convenience use that term we imply rather that the hypothesis was exposed to disconfirmation and was not disconfirmed. This point of view is compatible with all Humean philosophies of science which emphasize the impossibility of deductive proof for inductive laws. Recently Hanson (1958) and Popper (1959) have been particularly explicit upon this point. Many bodies of data collected in research on teaching have little or no probing value, and many hypothesis-sets are so double-jointed that they cannot be disconfirmed by available probes. We have no desire to increase the acceptability of such pseudo research. The research designs discussed below are believed to be sufficiently probing, however, to be well worth employing *where more efficient probes are unavailable.*

The notion that experiments never "confirm" theory, while correct, so goes against our attitudes and experiences as scientists as to be almost intolerable. Particularly does this emphasis seem unsatisfactory vis-à-vis the elegant and striking confirmations encountered in physics and chemistry, where the experimental data may fit in minute detail over numerous points of measurement a complex curve predicted by the theory. And the perspective becomes phenomenologically unacceptable to most of us when extended to the inductive achievements of vision. For example, it is hard to realize that the tables and chairs which we "see" before us are not "confirmed" or "proven" by the visual evidence, but are "merely" hypotheses about external objects not as yet disconfirmed by the multiple probes of the visual system. There is a grain of truth in these reluctances.

Varying degrees of "confirmation" are conferred upon a theory through the number of *plausible rival hypotheses* available to account for the data. The fewer such plausible rival hypotheses remaining, the greater the degree of "confirmation." Presumably, at any stage of accumulation of evidence, even for the most advanced science, there are numerous possible theories compatible with the data, particularly if all theories involving complex contingencies be allowed. Yet for "well-established" theories, and theories thoroughly probed by complex experiments, few if any rivals may be practically available or seriously proposed. This fewness is the epistemological counterpart of the positive affirmation of theory which elegant experiments seem to offer. A comparable fewness of rival hypotheses occurs in the phenomenally positive knowledge which vision seems to offer in contrast, for example, to the relative equivocality of blind tactile exploration.

In this perspective, the list of sources of invalidity which experimental designs control can be seen as a list of frequently plausible hypotheses which are rival to the hypothesis that the experimental variable has had an effect. Where an experimental design "controls" for one of these factors, it merely renders this rival hypothesis implausible, even though through possible complex coincidences it might still operate to produce the experimental outcome. The "plausible rival hypotheses" that have necessitated the routine use of special control groups have the status of well-established empirical laws: practice effects for adding a control group to Design 2, suggestibility for the placebo control group, surgical shock for the sham-operation control. Rival hypotheses are plausible insofar as we are willing to attribute to them the status of empirical laws. Where controls are lacking in a quasi-experiment, one must, in interpreting the results, consider in detail the likelihood of uncontrolled factors accounting for the results. The more implausible this becomes, the more "valid" the experiment.

As was pointed out in the discussion of the Solomon Four-Group Design 5, the more numerous and independent the ways in which the experimental effect is demonstrated, the less numerous and less plausible any singular rival invalidating hypothesis becomes. The appeal is to parsimony. The "validity" of the experiment becomes one of the relative credibility of rival theories: the theory that X had an effect versus the theories of causation involving the uncontrolled factors. If several sets of differences can all be explained by the single hypothesis that X has an effect, while several separate uncontrolled-variable effects must be hypothesized, a different one for each observed difference, then the effect of X becomes the most tenable. This mode of inference is frequently appealed to when scientists summarize a literature lacking in perfectly controlled experiments. Thus Watson (1959, p. 296) found the evidence for the deleterious effects of maternal deprivation confirmatory because it is supported by a wide variety of evidence-types, the specific inadequacies of which vary from study to study. Thus Glickman (1961), in spite of the presence of plausible rival hypotheses in each available study, found the evidence for a consolidation process impressive just because the plausible rival hypothesis is different from study to study. This inferential feature, commonly used in combining inferences from several studies, is deliberately introduced *within* certain quasi-experimental designs, especially in "patched-up" designs such as Design 15.

The appeal to parsimony is not deductively justifiable but is rather a general assumption about the nature of the world, underlying almost all use of theory in science, even though frequently erroneous in specific applications. Related to it is another plausibility argument which we will invoke perhaps most specifically with regard to the very widely used Design 10 (a good *quasi*-experimental design, often mistaken for the true Design 4). This is the assumption that, in cases of ignorance, a main effect of one variable is to be judged more likely than the interaction of two other variables; or, more generally, that main effects are more likely than interactions. In the extreme form, we can note that if every highest-order interaction is significant, if every effect is specific to certain values on all other potential treatment dimensions, then a science is not possible. If we are ever able to generalize, it is because the great bulk of potential determining factors can be disregarded. Underwood (1957b, p. 6) has referred to this as the assumption of finite causation. Elsewhere Underwood (1954) has tallied the frequency of main effects and interactions from the *Journal of Experimental Psychology,* confirming the relative rarity of significant interactions (although editorial selection favoring neat outcomes makes his finding suspect).

In what follows, we will first deal with single-group experiments. Since 1920 at least, the dominant experimental design in psychology and education has been a control group design, such as Design 4, Design 6, or perhaps most frequently Design 10, to be discussed later. In the social sciences and in thinking about field situations, the control group designs so dominate as to seem to many persons synonymous with experimentation. As a result, many research workers may give up attempting anything like experimentation in settings where control groups are not available and thus end up with more imprecision than is necessary. There are, in fact, several quasi-experimental designs applicable to single groups which might be used to advantage, with an experimental logic and interpretation, in many situations in which a control group design is impossible. Cooperation and experimental access often come in natural administrative units: a teacher has her own classroom available; a high school principal may be willing to introduce periodic morale surveys, etc. In such situations the differential treatment of segments within the administrative unit (required for the control group experiment) may be administratively impossible or, even if possible, experimentally undesirable owing to the reactive effects of arrangements. For these settings, single-group experiments might well be considered.

7. The Time-Series Experiment

The essence of the time-series design is the presence of a periodic measurement process on some group or individual and the introduction of an experimental change into this time series of measurements, the results of which are indicated by a discontinuity in the measurements recorded in the time series. It can be diagramed thus:

$$O_1 \; O_2 \; O_3 \; O_4 X O_5 \; O_6 \; O_7 \; O_8$$

This experimental design typified much of the classical nineteenth-century experimentation in the physical sciences and in biology. For example, if a bar of iron which has remained unchanged in weight for many months is dipped in a nitric acid bath and then removed, the inference tying together the nitric acid bath and the loss of weight by the iron bar would follow some such experimental logic. There may well have been "control groups" of iron bars remaining on the shelf that lost no weight, but the measurement and reporting of these weights would typically not be thought necessary or relevant. Thus it seems likely that this experimental design is frequently regarded as valid in the more successful sciences even though it rarely has accepted status in the enumerations of available experimental designs in the social sciences. (See, however, Maxwell, 1958; Underwood, 1957b, p. 133.) There are good reasons for this differential status and a careful consideration of them will provide a better understanding of the

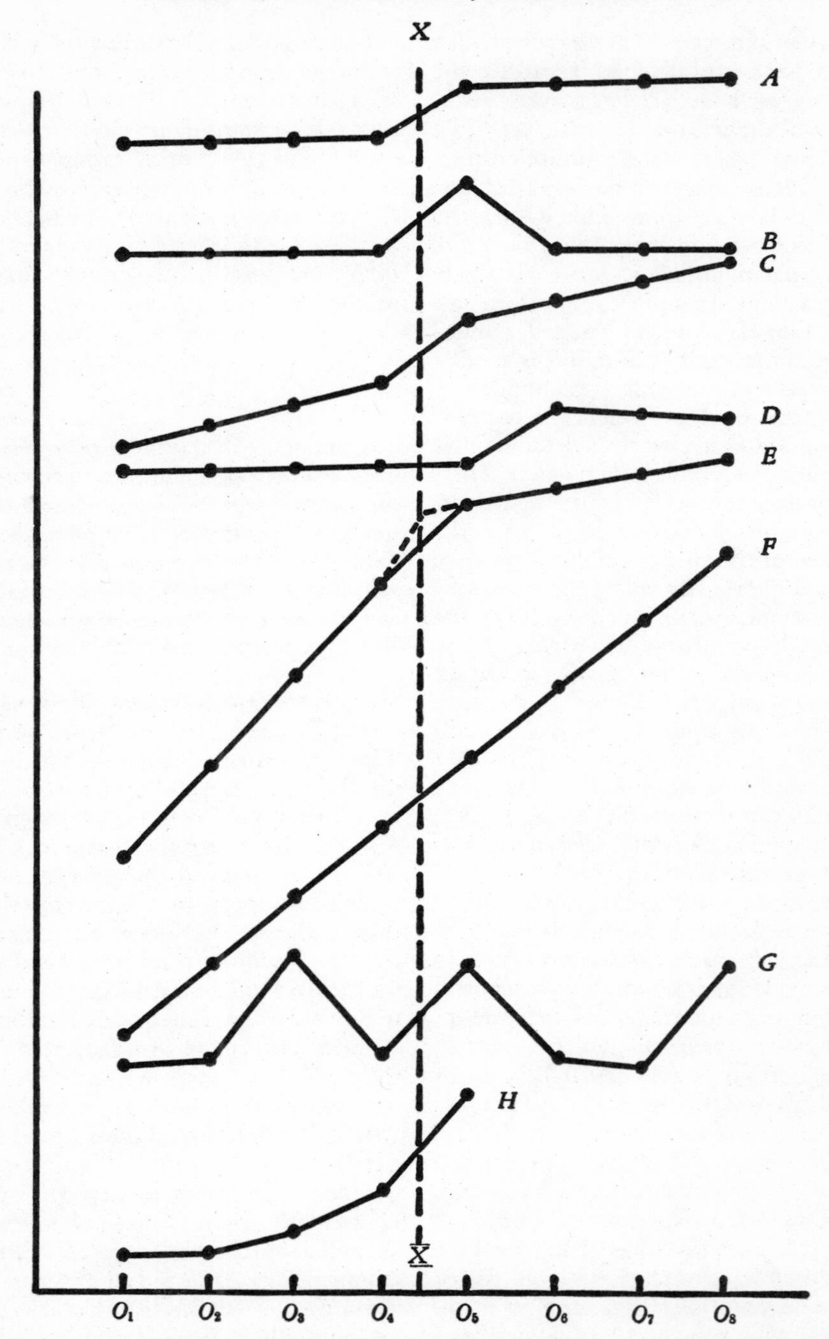

Fig. 3. Some Possible Outcome Patterns from the Introduction of an Experimental Variable at Point *X* into a Time Series of Measurements, O_1—O_8. Except for *D*, the O_4—O_5 gain is the same for all time series, while the legitimacy of inferring an effect varies widely, being strongest in *A* and *B*, and totally unjustified in *F*, *G*, and *H*.

conditions under which the design might meaningfully be employed by social scientists when more thorough experimental control is impossible. The design is typical of the classic experiments of the British Industrial Fatigue Research Board upon factors affecting factory outputs (e.g., Farmer, Brooks, & Chambers, 1923).

Figure 3 indicates some possible outcome patterns for time series into which an experimental alteration had been introduced as indicated by the vertical line *X*. For purposes of discussion let us assume that one will be tempted to infer that *X* had some effect in time series with outcomes such as *A* and *B* and possibly *C, D,* and *E,* but that one would not be tempted to infer an effect in time series such as *F, G,* and *H,* even were the jump in values from O_4 to O_5 as great and as statistically stable as were the O_4 to O_5 differences in *A* and *B,* for example. While discussion of the problem of statistical tests will be postponed for a few paragraphs, it is assumed that the problem of internal validity boils down to the question of plausible competing hypotheses that offer likely alternate explanations of the shift in the time series other than the effect of *X*. A tentative check-off of the controls provided by this experiment under these optimal conditions of outcome is provided in Table 2. The strengths of the time-series design are most apparent in contrast with Design 2, to which it has a superficial similarity in lacking a control group and in using before-and-after measures.

Scanning the list of problems of internal validity in Table 2, we see that failure to control history is the most definite weakness of Design 7. That is, the rival hypothesis exists that not *X* but some more or less simultaneous event produced the shift. It is upon the plausibility of ruling out such extraneous stimuli that credence in the interpretation of this experiment in any given instance must rest. Consider an experiment involving repeated measurements and the effect of a documentary film on students' optimism about the likelihood of war. Here the failure to provide a clear-cut control on *history* would seem very serious indeed since it is obvious that the students are exposed daily to many potentially relevant sources of stimulation beyond those under the experimenter's control in the classroom. Of course even here, were the experiment to be accompanied by a careful log of nonexperimental stimuli of possible relevance, plausible interpretation making the experiment worth doing might be possible. As has been noted above, the variable *history* is the counterpart of what in the physical and biological science laboratory has been called *experimental isolation*. The plausibility of *history* as an explanation for shifts such as those found in time-series *A* and *B* of Fig. 3 depends to a considerable extent upon the degree of experimental isolation which the experimenter can claim. Pavlov's conditioned-reflex studies with dogs, essentially "one-group" or "one-animal" experiments, would have been much less plausible as support of Pavlov's theories had they been conducted on a busy street corner rather than in a soundproof laboratory. What constitutes experimental isolation varies with the problem under study and the type of measuring device used. More precautions are needed to establish experimental isolation for a cloud chamber or scintillation counter study of subatomic particles than for the hypothetical experiment on the weight of bars of iron exposed to baths of nitric acid. In many situations in which Design 7 might be used, the experimenter could plausibly claim experimental isolation in the sense that he was aware of the possible rival events that might cause such a change and could plausibly discount the likelihood that they explained the effect.

Among other extraneous variables which might for convenience be put into *history* are the effects of weather and the effects of season. Experiments of this type are apt to extend over time periods that involve seasonal changes and, as in the studies of worker output, the seasonal fluctuations in illumination, weather, etc., may be confounded with the introduction of experimental change.

TABLE 2

SOURCES OF INVALIDITY FOR QUASI-EXPERIMENTAL DESIGNS 7 THROUGH 12

	Sources of Invalidity											
	Internal								External			
	History	Maturation	Testing	Instrumentation	Regression	Selection	Mortality	Interaction of Selection and Maturation, etc.	Interaction of Testing and X	Interaction of Selection and X	Reactive Arrangements	Multiple-X Interference
Quasi-Experimental Designs:												
7. Time Series $O\ O\ O\ OXO\ O\ O\ O$	−	+	+	?	+	+	+	+	−	?	?	
8. Equivalent Time Samples Design $X_1O\ X_0O\ X_1O\ X_0O$, etc.	+	+	+	+	+	+	+	+	−	?	−	−
9. Equivalent Materials Samples Design $M_aX_1O\ M_bX_0O\ M_cX_1O\ M_dX_0O$, etc.	+	+	+	+	+	+	+	+	−	?	?	−
10. Nonequivalent Control Group Design $O\ X\ O$ - - - - - - - - - $O\ \ \ \ O$	+	+	+	+	?	+	+	−	−	?	?	
11. Counterbalanced Designs $X_1O\ X_2O\ X_3O\ X_4O$ - - - - - - - - - $X_2O\ X_4O\ X_1O\ X_3O$ - - - - - - - - - $X_3O\ X_1O\ X_4O\ X_2O$ - - - - - - - - - $X_4O\ X_3O\ X_2O\ X_1O$	+	+	+	+	+	+	+	?	?	?	?	−
12. Separate-Sample Pretest-Posttest Design $R\ O\ (X)$ $R\ \ \ \ X\ O$	−	−	+	?	+	+	−	−	+	+	+	
12*a*. $R\ O\ (X)$ $R\ \ \ \ X\ O$ - - - - - - - - - $R\ \ \ \ O\ (X)$ $R\ \ \ \ \ \ \ X\ O$	+	−	+	?	+	+	−	+	+	+	+	
12*b*. $R\ O_1\ \ \ (X)$ $R\ \ \ O_2\ (X)$ $R\ \ \ \ \ \ X\ \ O_3$	−	+	+	?	+	+	−	?	+	+	+	
12*c*. $R\ O_1\ X\ O_2$ $R\ \ \ \ \ X\ O_3$	−	−	+	?	+	+	+	−	+	+	+	

Perhaps best also included under *history*, although in some sense akin to *maturation*, would be periodical shifts in the time series related to institutional customs of the group such as the weekly work-cycles, pay-period cycles, examination periods, vacations, and student festivals. The observational series should be arranged so as to hold known cycles constant, or else be long enough to include several such cycles in their entirety.

To continue with the factors to be controlled: *maturation* seems ruled out on the grounds that if the outcome is like those in illustrations *A* and *B* of Fig. 3, maturation does not usually provide plausible rival hypotheses to explain a shift occurring between O_4 and O_5 which did not occur in the previous time periods under observation. (However, maturation may not always be of a smooth, regular nature. Note how the abrupt occurrence of menarche in first-year junior high school girls might in a Design 7 appear as an effect of the shift of schools upon physiology records, did we not know better.) Similarly, testing seems, in general, an implausible rival hypothesis for a jump between O_4 and O_5. Had one only the observations at O_4 and O_5, as in Design 2, this means of rendering maturation and test-retest effects implausible would be lacking. Herein lies the great advantage of this design over Design 2.

In a similar way, many hypotheses invoking changes in *instrumentation* would lack a specific rationale for expecting the instrument error to occur on this particular occasion, as opposed to earlier ones. However, the question mark in Table 2 calls attention to situations in which a change in the calibration of the measurement device could be misinterpreted as the effect of *X*. If the measurement procedure involves the judgments of human observers who are aware of the experimental plan, pseudo confirmation of the hypothesis can occur as a result of the observer's expectations. Thus, the experimental change of putting into office a new principal may produce a change in the recording of discipline infractions rather than in the infraction rate itself. Design 7 may frequently be employed to measure effects of a major change in administrative policy. Bearing this in mind, one would be wise to avoid shifting measuring instruments at the same time he shifts policy. In most instances, to preserve the interpretability of a time series, it would be better to continue to use a somewhat antiquated device rather than to shift to a new instrument.

Regression effects are usually a negatively accelerated function of elapsed time and are therefore implausible as explanations of an effect at O_5 greater than the effects at O_2, O_3, and O_4. *Selection* as a source of main effects is ruled out in both this design and in Design 2, if the same specific persons are involved at all *O*s. If data from a group is basically collected in terms of individual group members, then mortality may be ruled out in this experiment as in Design 2. However, if the observations consist of collective products, then a record of the occurrence of absenteeism, quitting, and replacement should be made to insure that coincidences of personnel change do not provide plausible rival hypotheses.

Regarding external validity, it is clear that the experimental effect might well be specific to those populations subject to repeated testing. This is hardly likely to be a limitation in research on teaching in schools, unless the experiment is conducted with artificial *O*s not common to the usual school setting. Furthermore, this design is particularly appropriate to those institutional settings in which records are regularly kept and thus constitute a natural part of the environment. Annual achievement tests in the public schools, illness records, etc., usually are nonreactive in the sense that they are typical of the universe to which one wants to generalize. The *selection-X* interaction refers to the limitation of the effects of the experimental variable to that specific sample and to the possibility that this reaction would not be typical of some more general universe of interest for which the naturally aggregated exposure-group was a biased sample. For example, the data requirements may limit one to those students who have had perfect attendance records over long periods, an obviously select subset. Further, if novel *O*s have been used, this repetitive occurrence may have provoked absenteeism.

If such time series are to be interpreted as experiments, it seems essential that the experimenter must specify in advance the expected time relationship between the introduction of the experimental variable and the manifestation of an effect. If this had been done, the pattern indicated in time-series *D* of Fig. 3 could be almost as definitive as that in *A*. Exploratory surveys opportunistically deciding upon interpretations of delayed effect would require cross-validation before being interpretable. As the time interval between *X* and effect increases, the plausibility of effects from extraneous historical events also increases.

It also seems imperative that the *X* be specified before examining the outcome of the time series. The post hoc examination of a time series to infer what *X* preceded the most dramatic shift must be ruled out on the grounds that the opportunistic capitalization on chance which it allows makes any approach to testing the significance of effects difficult if not impossible.

The prevalence of this design in the more successful sciences should give us some respect for it, yet we should remember that the facts of "experimental isolation" and "constant conditions" make it more interpretable for them than for us. It should also be remembered that, in their use of it, a single experiment is never conclusive. While a control group may never be used, Design 7 is repeated in many different places by various researchers before a principle is established. This, too, should be our use of it. *Where nothing better controlled is possible,* we will use it. We will organize our institutional bookkeeping to provide as many time series as possible for such evaluations and will try to examine in more detail than we have previously the effects of administrative changes and other abrupt and arbitrary events as *X*s. But these will not be regarded as definitive until frequently replicated in various settings.

Tests of Significance for the Times-Series Design

If the more advanced sciences use tests of significance less than do psychology and education, it is undoubtedly because the magnitude and the clarity of the effects with which they deal are such as to render tests of significance unnecessary. If our conventional tests of significance were applied, high degrees of significance would be found. It seems typical of the ecology of the social sciences, however, that they must work the low-grade ore in which tests of significance are necessary. It also seems likely that wherever common sense or intuitive considerations point to a clear-cut effect, some test of significance that formalizes considerations underlying the intuitive judgment is usually possible. Thus tests of significance of the effects of *X* that would distinguish between the several outcomes illustrated in Fig. 3, judging *A* and *B* to be significant and *F* and *G* not significant, may be available. We shall discuss a few possible approaches.

First, however, let us reject certain conceivable approaches as inadequate. If the data in Fig. 3 represent group means, then a simple significance test of the difference between the observations of O_4 and O_5 is insufficient. Even if in series *F* and *G*, these provided *t* ratios that were highly significant, we would not find the data evidence of effect of *X* because of the presence of other similar significant shifts occurring on occasions for which we had no matching experimental explanation. Where one is dealing with the kind of data provided in national opinion surveys, it is common to encounter highly significant shifts from one survey to the next which are random noise from the point of view of the interpreting scientist, inasmuch as they represent a part of the variation in the phenomena for which he has no explanation. The effect of a clear-cut event or experimental variable must rise above this ordinary level of shift in order to be interpretable. Similarly, a test of significance involving the pooled data for all of the pre-*X* and post-*X* observations is inadequate, inasmuch as it would not distinguish between instances of type *F* and instances of type *A*.

There is a troublesome nonindependence involved which must be considered in developing a test of significance. Were such nonindependence homogeneously distributed across all observations, it would be no threat to internal validity, although a limitation to external validity. What is troublesome is that in almost every time series it will be found that adjacent observations are more similar than nonadjacent ones (i.e., that the autocorrelation of lag 1 is greater than that for lag 2, etc.). Thus, an extraneous influence or random disturbance affecting an observation point at, say, O_5 or O_6, will also disturb O_7 and O_8, so that it is illegitimate to treat them as several independent departures from the extrapolation of the O_1—O_4 trend.

The test of significance employed will, in part, depend upon the hypothesized nature of the effect of X. If a model such as line B is involved, then a test of the departure of O_5 from the extrapolation of O_1—O_4 could be used. Mood (1950, pp. 297–298) provides such a test. Such a test could be used for all instances, but it would seem to be unnecessarily weak where a continuous improvement, or increased rate of gain, were hypothesized. For such cases, a test making use of all points would seem more appropriate. There are two components which might enter into such tests of significance. These are intercept and slope. By intercept we refer to the jump in the time series at X (or at some specified lag after X). Thus lines A and C show an intercept shift with no change in slope. Line E shows a change in slope but no change in intercept in that the pre-X extrapolation to X and the post-X extrapolation to X coincide. Often both intercept, and slope would be changed by an effective X. A pure test of intercept might be achieved in a manner analogous to working the Mood test from both directions at once. In this case, two extrapolated points would be involved, with both pre-X and post-X observations being extrapolated to a point X halfway between O_4 and O_5.

Statistical tests would probably involve, in all but the most extended time series, linear fits to the data, both for convenience and because more exact fitting would exhaust the degrees of freedom, leaving no opportunity to test the hypothesis of change. Yet frequently the assumption of linearity may not be appropriate. The plausibility of inferring an effect of X is greatest adjacent to X. The more gradual or delayed the supposed effect, the more serious the confound with history, because the possible extraneous causes become more numerous.

8. The Equivalent Time-Samples Design

The most usual form of experimental design employs an equivalent sample of persons to provide a baseline against which to compare the effects of the experimental variable. In contrast, a recurrent form of one-group experimentation employs two equivalent samples of occasions, in one of which the experimental variable is present and in another of which it is absent. This design can be diagramed as follows (although a random rather than a regular alternation is intended):

$$X_1O \quad X_0O \quad X_1O \quad X_0O$$

This design can be seen as a form of the time-series experiment with the repeated introduction of the experimental variable. The experiment is most obviously useful where the effect of the experimental variable is anticipated to be of transient or reversible character. While the logic of the experiment may be seen as an extension of the time-series experiment, the mode of statistical analysis is more typically similar to that of the two-group experiment in which the significance of the difference between the means of two sets of measures is employed. Usually the measurements are quite specifically paired with the presentations of the experimental variable, frequently being concomitant, as in studies of learning, work production, conditioning, physiological reaction, etc. Perhaps the most typical early use of this experimental design, as in the studies of efficiency of students' work under various conditions by Allport (1920) and Sorokin (1930), involved the comparison of two experimental variables with each other, i.e., X_1 versus X_2 rather than one with a control. For most purposes, the simple alternation of conditions and the employment of a consistent time spacing are undesirable, particularly when they may introduce confounding with a daily, weekly, or monthly cycle, or when through the predictable periodicity an unwanted conditioning to the temporal interval may accentuate the difference between one presentation and another. Thus Sorokin made sure that each experimental treatment occurred equally often in the afternoon and the forenoon.

Most experiments employing this design have used relatively few repetitions of each experimental condition, but the type of extension of sampling theory represented by Brunswik (1956) calls attention to the need for large, representative, and equivalent random samplings of time periods. Kerr (1945) has perhaps most nearly approximated this ideal in his experiments on the effects of music upon industrial production. Each of his several experiments involved a single experimental group with a randomized, equivalent sample of days over periods of months. Thus, in one experiment he was able to compare 56 music days with 51 days without music, and in another he was able to compare three different types of music, each represented by equivalent samples of 14 days.

As employed by Kerr, for example, Design 8 seems altogether internally valid. *History,* the major weakness of the time-series experiment, is controlled by presenting X on numerous separate occasions, rendering extremely unlikely any rival explanation based on the coincidence of extraneous events. The other sources of invalidity are controlled by the same logic detailed for Design 7. With regard to external validity, generalization is obviously possible only to frequently tested populations. The reactive effect of arrangements, the awareness of experimentation, represents a particular vulnerability of this experiment. Where separate groups are getting the separate Xs, it is possible (particularly under Design 6) to have them totally unaware of the presence of an experiment or of the treatments being compared. This is not so when a single group is involved, and when it is repeatedly being exposed to one condition or another, e.g., to one basis for computing payment versus another in Sorokin's experiment; to one condition of work versus another in Allport's; to one kind of ventilation versus another in Wyatt, Fraser, and Stock's (1926) studies; and to one kind of music versus another in Kerr's (although Kerr took elaborate precautions to make varied programing become a natural part of the working environment). As to the interaction of *selection* and X*:* there is as usual the limitation of the generalization of the demonstrated effects of X to the particular type of population involved.

This experimental design carries a hazard to external validity which will be found in all of those experiments in this paper in which multiple levels of X are presented to the *same* set of persons. This effect has been labeled "multiple-X interference." The effect of X_1, in the simplest situation in which it is being compared with X_0, can be generalized only to conditions of repetitious and spaced presentations of X_1. No sound basis is provided for generalization to possible situations in which X_1 is continually present, or to the condition in which it is introduced once and once only. In addition, the X_0 condition or the absence of X is not typical of periods without X in general, but is only representative of absences of X interspersed among presences. If X_1 has some extended effect carrying over into the non-X periods, as usually would seem likely, the experimental design may underestimate the effect of X_1 as compared with a Design 6 study, for example. On the other hand, the very fact of frequent shifts may increase the stimulus value of an X over what it would be under a continuous, homogeneous presentation. Hawaiian music in Kerr's study might affect work quite differently when interspersed for a day among days of other music than it would as a continuous diet. Ebbinghaus' (1885) experimental designs may be regarded as essentially of this type and, as Underwood (1957a) has pointed out, the laws which he found are limited in their generalizability to a population of persons who have learned dozens of other highly similar lists. Many of his findings do not in fact hold for persons learning a single list of nonsense syllables. Thus, while the design is internally valid, its external validity may be seriously limited for some types of content. (See also Kempthorne, 1952, Ch. 29.)

Note, however, that many aspects of teaching on which one would like to experiment may very well have effects limited for all practical purposes to the period of actual presence of X. For such purposes, this design might be quite valuable. Suppose a teacher questions the value of oral recitation versus individual silent study. By varying these two procedures over a series of lesson units, one could arrange an interpretable experiment. The effect of the presence of a parent-observer in the classroom upon students' volunteer discussion could be studied in this way. Awareness of such designs can place an experimental testing of alternatives within the grasp of an individual teacher. This could pilot-test procedures which if promising might be examined by larger, more coordinated experiments.

This approach could be applied to a sampling of occasions for a single individual. While tests of significance are not typically applied, this is a recurrent design in physiological research, in which a stimulus is repeatedly applied to one animal, with care taken to avoid any periodicity in the stimulation, the latter feature corresponding to the randomization requirement for occasions demanded by the logic of the design. Latin squares rather than simple randomization may also be used (e.g., Cox, 1951; Maxwell, 1958).

Tests of Significance for Design 8

Once again, we need appropriate tests of significance for this particular type of design. Note that two dimensions of generalization are implied: generalization across occasions and generalization across persons. If we consider an instance in which only one person is employed, the test of significance will obviously be limited to generalizations about this particular person and will involve a generalization across instances, for which purpose it will be appropriate to use a t with degrees of freedom equal to the number of occasions less two. If one has individual records on a number of persons undergoing the same treatment, all a part of the same group, then data are available also for generalization across persons. In this usual situation two strategies seem common. A wrong one is to generate for each individual a single score for each experimental treatment, and then to employ tests of significance of the difference between means with correlated data. While tests of significance were not actually employed, this is the logic of Allport's and Sorokin's analyses. But where only one or two repetitions of each experimental condition are involved, sampling errors of occasions may be very large or the control of history may be very poor. Chance sampling errors of occasions could contribute what would appear under this analysis to be significant differences among treatments. This seems to be a very serious error if the effect of occasions is significant and appreciable. One could, for example, on this logic get a highly significant difference between X_1 and X_2 where each has been presented only once and where on one occasion some extraneous event had by chance produced a marked result. It seems essential therefore that at least two occasions be "nested" within each treatment and that degrees of freedom between occasions within treatments be represented. This need is probably most easily met by initially testing the difference between treatment means against a between-occasions-within-treatments error term. After the significance of the treatment effect has been established in this way, one could proceed to find for what proportion of the subjects it held, and thus obtain evidence relevant to the generalizability of the effect across persons. Repeated measurements and sampling of occasions pose many statistical problems, some of them still unresolved (Collier, 1960; Cox, 1951; Kempthorne, 1952).

9. The Equivalent Materials Design

Closely allied to the equivalent time-samples design is Design 9, basing its argument on the equivalence of samples of materials to which the experimental variables being compared are applied. Always or almost always, equivalent time samples are also involved, but they may be so finely or intricately interspersed that there is practical temporal equivalence. In a one-group repeated-X design, equivalent materials are required whenever the nature of the experimental variables is such that the effects are enduring and the different treatments and repeats of treatments must be applied to nonidentical content. The design may be indicated in this fashion:

$$M_aX_1O \quad M_bX_0O \quad M_cX_1O \quad M_dX_0O \quad \text{etc.}$$

The Ms indicate specific materials, the sample M_a, M_c, etc., being, in sampling terms, equal to the sample M_b, M_d, etc. The importance of the sampling equivalence of the two sets of materials is perhaps better indicated if the design is diagramed in this fashion:

$$\text{one person or group} \begin{cases} \text{Materials Sample } A \ (O) \ X_0 \ O \\ \text{Materials Sample } B \ (O) \ X_1 \ O \end{cases}$$

The Os in parentheses indicate that in some designs a pretest will be used and in others not.

Jost's (1897) early experiment on massed versus distributed practice provides an excellent illustration. In his third experiment, 12 more or less randomly assembled lists of 12 nonsense syllables each were prepared. Six of the lists were assigned to distributed practice and six to massed practice. These 12 were then simultaneously learned over a seven-day period, their scheduling carefully intertwined so as to control for fatigue, etc. Seven such sets of six distributed and six massed lists were learned over a period lasting from November 6, 1895, to April 7, 1896. In the end, Jost had results on 40 different nonsense syllable lists learned under massed practice and 40 learned under distributed practice. The interpretability of the differences found on the one subject, Professor G. E. Müller, depends upon the sampling equivalence of the nonidentical lists involved. Within these limits, this experiment seems to have internal validity. The findings are of course restricted to the psychology of Professor G. E. Müller in 1895 and 1896 and to the universe of memory materials sampled. To enable one to generalize across persons in achieving a more general psychology, replication of the experiment on numerous persons is of course required.

Another illustration comes from early studies of conformity to group opinion. For example, Moore (1921) obtained a "control" estimate of retest stability of questionnaire responses from one set of items, and then compared this with the change resulting when, with another set of items, the retest was accompanied by a statement of majority opinion. Or consider a study in which students are asked to express their opinions on a number of issues presented in a long questionnaire. These questions are then divided into two groups as equivalent as possible. At a later time the questionnaires are handed back to the students and the group vote for each item indicated. These votes are falsified, to indicate majorities in opposite directions for the two samples of items. As a post-X measure, the students are asked to vote again on all items. Depending upon the adequacy of the argument of sampling equivalence of the two sets of items, the differences in shifts between the two experimental treatments would seem to provide a definitive experimental demonstration of the effects of the reporting of group opinions, even in the absence of any control group of persons.

Like Design 8, Design 9 has internal validity on all points, and in general for the same reasons. We may note, with regard to external validity, that the effects in Design 9, like those in all experiments involving repeated measures, may be quite specific to persons repeatedly measured. In learning experiments, the measures are so much a part of the experimental setting in the typical method used today (although not necessarily in Jost's method, in which the practices involved controlled numbers of readings of the lists) that this limitation on generalization becomes irrelevant. Reactive arrangements seem to be less certainly involved in Design 9 than in Design 8 because of the heterogeneity of the materials and the greater possibility that the subjects will not be aware that they are getting different treatments at different times for different items. This low reactivity would not be found in Jost's experiment but it would be found in the conformity study. Interference among the levels of the experimental variable or interference among the materials seems likely to be a definite weakness for this experiment, as it is for Design 8.

We have a specific illustration of the kind of limitation thus introduced with regard to Jost's findings. He reported that spaced learning was more efficient than massed practice. From the conditions of his experimentation in general, we can see that he was justified in generalizing only to persons who were learning many lists, that is, persons for whom the general interference level was high. Contemporary research indicates that the superiority of spaced learning is limited to just such populations, and that for persons learning highly novel materials for the first time, no such advantage is present (Underwood & Richardson, 1958).

Statistics for Design 9

The sampling of materials is obviously relevant to the validity and the degree of proof of the experiment. As such, the N for the computation of the significance of the differences between the means of treatment groups should probably have been an N of lists in the Jost experiment (or an N of items in the conformity study) so as to represent this relevant sampling domain. This must be supplemented by a basis for generalizing across persons. Probably the best practice at the present time is to do these seriatim, establishing the generalization across the sample of lists or items first, and then computing an experimental effects score for each particular person and employing this as a basis for generalizing across persons. (Note the cautionary literature cited above for Design 8.)

10. The Nonequivalent Control Group Design

One of the most widespread experimental designs in educational research involves an experimental group and a control group both given a pretest and a posttest, but in which the control group and the experimental group do not have pre-experimental sampling equivalence. Rather, the groups constitute naturally assembled collectives such as classrooms, as similar as availability permits but yet not so similar that one can dispense with the pretest. The assignment of X to one group or the other is assumed to be random and under the experimenter's control.

$$\frac{O \quad X \quad O}{O \qquad\quad O}$$

Two things need to be kept clear about this design: First, it is not to be confused with Design 4, the Pretest-Posttest Control Group Design, in which experimental subjects are assigned *randomly* from a common population to the experimental and the control group. Second, in spite of this, Design 10 should be recognized as well worth using in many instances in which Designs 4, 5, or 6 are impossible. In particular it should be recognized that the addition of even an unmatched or nonequivalent control group reduces greatly the equivocality of interpretation over what is obtained in Design 2, the One-Group Pretest-Posttest Design. The more similar the experimental and the control groups are in their recruitment, and the more this similarity is confirmed by the scores on the pretest, the more effective this control becomes. Assuming that these desiderata are approximated for purposes of internal validity, we can regard the design as controlling the main effects of history, maturation, testing, and instrumentation, in that the difference for the experimental group between pretest and posttest (if greater than that for the control group) cannot be explained by main effects of these variables such as would be found affecting both the experimental and the control group. (The cautions about intrasession history noted for Design 4 should, however, be taken very seriously.)

An effort to explain away a pretest-posttest gain specific to the experimental group in terms of such extraneous factors as history, maturation, or testing must hypothesize an interaction between these variables and the specific selection differences that distinguish the experimental and control groups. While in general such interactions are unlikely, there are a number of situations in which they might be invoked. Perhaps most common are interactions involving *maturation*. If the experimental group consists of psychotherapy patients and the control group some other handy population tested and retested, a gain specific to the experimental group might well be interpreted as a spontaneous remission process specific to such an extreme group, a gain that would have occurred even without X. Such a selection-maturation interaction (or a selection-history interaction, or a selection-testing interaction) could be mistaken for the effect of X, and thus represents a threat to the *internal* validity of the experiment. This possibility has been represented in the eighth column of Table 2 and is the main factor of *internal* validity which distinguishes Designs 4 and 10.

A concrete illustration from educational research may make this point clear. Sanford and Hemphill's (1952) study of the effects of a psychology course at Annapolis provides an excellent illustration of Design 10. In this study, the Second Class at Annapolis provided the experimental group and the Third Class the control group. The greater gains for the experimental group might be explained away as a part of some general sophistication process occurring maximally in the first two classes and only in minimal degree in the Third and Fourth, thus representing an interaction between the selection factors differentiating the experimental and control groups and natural changes (maturation) characteristic of these groups, rather than any effect of the experimental program. The particular control group utilized by Sanford and Hemphill makes possible some check on this rival interpretation (somewhat in the manner of Design 15 below). The selection-maturation hypothesis would predict that the Third Class (control group) in its initial test would show a superiority to the pretest measures for the Second Class (experimental group) of roughly the same magnitude as that found between the experimental group pretest and posttest. Fortunately for the interpretation of their experiment, this was not generally so. The class differences on the pretest were in most instances not in the same direction nor of the same magnitude as the pretest-posttest gains for the experimental group. However, their finding of a significant gain for the experimental group in confidence scores on the social situations questionnaire can be explained away as a selection-maturation artifact. The experimental group shows a gain from 43.26 to 51.42, whereas the Third Class starts out with a score of 55.82 and goes on to a score of 56.78.

The hypothesis of an interaction between selection and maturation will occasionally be tenable even where the groups are identical in pretest scores. The commonest of these instances will be where one group has a higher rate of maturation or autonomous change than the other. Design 14 offers an extension of 10 which would tend to rule this out.

Regression provides the other major internal validity problem for Design 10. As indicated by the "?" in Table 2, this hazard is avoidable but one which is perhaps more frequently tripped over than avoided. In general, if either of the comparison groups has been selected for its extreme scores on *O* or correlated measures, then a difference in degree of shift from pretest to posttest between the two groups may well be a product of regression rather than the effect of *X*. This possibility has been made more prevalent by a stubborn, misleading tradition in educational experimentation, in which matching has been regarded as the appropriate and sufficient procedure for establishing the pre-experimental equivalence of groups. This error has been accompanied by a failure to distinguish Designs 4 and 10 and the quite different roles of matching on pretest scores under the two conditions. In Design 4, matching can be recognized as a useful adjunct to randomization but not as a substitute for it: in terms of scores on the pretest or on related variables, the total population available for experimental purposes can be organized into carefully matched pairs of subjects; members of these pairs can then be assigned *at random* to the experimental or the control conditions. Such matching plus subsequent randomization usually produces an experimental design with greater precision than would randomization alone.

Not to be confused with this ideal is the procedure under Design 10 of attempting to compensate for the differences between the nonequivalent experimental and control groups by a procedure of matching, when random assignment to treatments is not possible. If in Design 10 the means of the groups are substantially different, then the process of matching not only fails to provide the intended equation but in addition insures the occurrence of unwanted regression effects. It becomes predictably certain that the two groups will differ on their posttest scores altogether independently of any effects of *X*, and that this difference will vary directly with the difference between the total populations from which the selection was made and inversely with the test-retest correlation. Rulon (1941), Stanley and Beeman (1958), and R. L. Thorndike (1942) have discussed this problem thoroughly and have called attention to covariance analysis and to other statistical techniques suggested by Johnson and Neyman (see Johnson & Jackson, 1959, pp. 424–444) and by Peters and Van Voorhis (1940) for testing the effects of the experimental variable without the procedure of matching. Recent cautions by Lord (1960) concerning the analysis of covariance when the covariate is not perfectly reliable should be considered, however. Simple gain scores are also applicable but usually less desirable than analysis of covariance. Application of analysis of covariance to this Design 10 setting involves assumptions (such as that of homogeneity of regression) less plausible here than in Design 4 settings (Lindquist, 1953).

In interpreting published studies of Design 10 in which matching was used, it can be noted that the direction of error is predictable. Consider a psychotherapy experiment using ratings of dissatisfaction with one's own personality as *O*. Suppose the experimental group consists of therapy applicants and the matched control group of "normal" persons. Then the control group will turn out to represent extreme low scores from the normal group (selected because of their extremity), will regress on the posttest in the direction of the normal group average, and thus will make it less likely that a significant effect of therapy can be shown, rather than produce a spurious impression of efficacy for the therapeutic procedure.

The illustration of psychotherapy applicants also provides an instance in which the assumptions of homogeneous regression and of sampling from the same universe, except for extremity of scores, would seem likely to be inappropriate. The inclusion of normal controls in psychotherapy research is of some use, but extreme caution must be employed in interpreting results. It seems important to distinguish two versions of Design 10, and to give them different status as approximations of true experimentation. On the one hand, there is the situation in which the experimenter has two natural groups available, e.g., two classrooms, and has free choice in deciding which gets *X*, or at least has no reason to suspect differential recruitment related to *X*. Even though the groups may differ in initial means on *O*, the study may approach true experimentation. On the other hand, there are instances of Design 10 in which the respondents clearly are self-selected, the experimental group having deliberately sought out exposure to *X*, with no control group available from this same population of seekers. In this latter case, the assumption of uniform regression between experimental and control groups becomes less likely, and selection-maturation interaction (and the other selection interactions) become more probable. The "self-selected" Design 10 is thus much weaker, but it does provide information which in many instances would rule out the hypothesis that *X* has an effect. The control group, even if widely divergent in method of recruitment and in mean level, assists in the interpretation.

The threat of testing to external validity is as presented for Design 4 (see page 188). The question mark for interaction of selection and *X* reminds us that the effect of *X* may well be specific to respondents selected as the ones in our experiment have been. Since the requirements of Design 10 are likely to put fewer limitations on our freedom to sample widely than do those of Design 4, this specificity will usually be less than it would be for a laboratory experiment. The threat to external validity represented by reactive arrangements is present, but probably to a lesser degree than in most true experiments, such as Design 4.

Where one has the alternative of using two intact classrooms with Design 10, or taking random samples of the students out of the classrooms for different experimental treatments under a Design 4, 5, or 6, the latter arrangement is almost certain to be the more reactive, creating more awareness of experiment, I'm-a-guinea-pig attitude, and the like.

The Thorndike studies of formal discipline and transfer (e.g., E. L. Thorndike & Woodworth, 1901; Brolyer, Thorndike, & Woodyard, 1927) represent applications of Design 10 to *X*s uncontrolled by the experimenter. These studies avoided in part, at least, the mistake of regression effects due to simple matching, but should be carefully scrutinized in terms of modern methods. The use of covariance statistics would probably have produced stronger evidence of transfer from Latin to English vocabulary, for example.

In the other direction, the usually positive, albeit small, transfer effects found could be explained away not as transfer but as the selection into Latin courses of those students whose annual rate of vocabulary growth would have been greater than that of the control group even without the presence of the Latin instruction. This would be classified here as a selection-maturation interaction. In many school systems, this rival hypothesis could be checked by extending the range of pre-Latin *O*s considered, as in a Design 14. These studies were monumental efforts to get experimental thinking into field research. They deserve renewed attention and extension with modern methods.

11. Counterbalanced Designs

Under this heading come all of those designs in which experimental control is achieved or precision enhanced by entering all respondents (or settings) into all treatments. Such designs have been called "rotation experiments" by McCall (1923), "counterbalanced designs" (e.g., Underwood, 1949), cross-over designs (e.g., Cochran & Cox, 1957; Cox, 1958), and switch-over designs (Kempthorne, 1952). The Latin-square arrangement is typically employed in the counterbalancing. Such a Latin square is employed in Design 11, diagramed here as a quasi-experimental design, in which four experimental treatments are applied in a restrictively *randomized* manner in turn to four naturally assembled groups or even to four individuals (e.g., Maxwell, 1958):

	Time 1	*Time 2*	*Time 3*	*Time 4*
Group A	X_1O	X_2O	X_3O	X_4O
Group B	X_2O	X_4O	X_1O	X_3O
Group C	X_3O	X_1O	X_4O	X_2O
Group D	X_4O	X_3O	X_2O	X_1O

The design has been diagramed with posttests only, because it would be especially preferred where pretests were inappropriate, and designs like Design 10 were unavailable. The design contains three classifications (groups, occasions, and *X*s or experimental treatments). Each classification is "orthogonal" to the other two in that each variate of each classification occurs equally often (once for a Latin square) with each variate of each of the other classifications. To begin with, it can be noted that each treatment (each *X*) occurs once and only once in each column and only once in each row. The same Latin square can be turned so that *X*s become row or column heads, e.g.:

	X_1	X_2	X_3	X_4
Group A	t_1O	t_2O	t_3O	t_4O
Group B	t_3O	t_1O	t_4O	t_2O
Group C	t_2O	t_4O	t_1O	t_3O
Group D	t_4O	t_3O	t_2O	t_1O

Sums of scores by *X*s thus are comparable in having each time and each group represented in each. The differences in such sums could not be interpreted simply as artifacts of the initial group differences or of practice effects, history, etc. Similarly comparable are the sums of the rows for intrinsic group differences, and the sums of the columns of the first presentation for the differences in occasions. In analysis of variance terms, the design thus appears to provide data on three main effects in a design with the number of cells usually required for two. Thinking in analysis of variance terms makes apparent the cost of this greater efficiency: What appears to be a significant main effect for any one of the three classification criteria could be instead a significant interaction of a complex form between the other two (Lindquist, 1953, pp. 258–264). The apparent differences among the effects of the *X*s could instead be a specific complex interaction effect between the group differences and the occasions. Inferences as to effects of *X* will be dependent upon the plausibility of this rival hypothesis, and will therefore be discussed in more detail.

First, let us note that the hypothesis of such interaction is more plausible for the quasi-experimental application described than for the applications of Latin squares in the true experiments described in texts covering the topic. In what has been described as the dimension of groups, two possible sources of systematic effects are confounded. First, there are the systematic selection factors involved in the natural assemblage of the groups. These factors can be expected both to have main effects and to interact with history, maturation, practice effects, etc. Were a fully controlled experiment to have been organized in this way, each person would have been assigned to each group independently and at random, and this source of both main and interaction effects would have been removed, at least to the extent of sampling error. It is characteristic of the quasi-experiment that the counterbalancing was introduced to provide a kind of equation just because such random assignment was not possible. (In contrast, in fully controlled experiments, the Latin square is employed for reasons of economy or to handle problems specific to the sampling of land parcels.) A second possible source of effects confounded with groups is that associated with specific sequences of treatments. Were all replications in a true experiment to have followed the same Latin square, this source of main and interaction effects would also have been present. In the typical *true* experiment, however, some replication sets of respondents would have been assigned different specific Latin squares, and the systematic effect of specific sequences eliminated. This also rules out the possibility that a specific systematic interaction has produced an apparent main effect of *X*s.

Occasions are likely to produce a main effect due to repeated testing, maturation, practice, and cumulative carry-overs, or transfer. History is likewise apt to produce effects for occasions. The Latin-square arrangement, of course, keeps these main effects from contaminating the main effects of *X*s. But where main effects symptomatize significant heterogeneity, one is probably more justified in suspecting significant interactions than when main effects are absent. Practice effects, for example, may be monotonic but are probably nonlinear, and would generate both main and interaction effects. Many uses of Latin squares in true experiments, as in agriculture, for instance, do not involve repeated measurements and do not typically produce any corresponding systematic column effects. Those of the cross-over type, however, share this potential weakness with the quasi-experiments.

These considerations make clear the extreme importance of replication of the quasi-experimental design with different specific Latin squares. Such replications in sufficient numbers would change the quasi-experiment into a true experiment. They would probably also involve sufficient numbers of groups to make possible the random assignment of intact groups to treatments, usually a preferable means of control. Yet, lacking such possibilities, a single Latin square represents an intuitively satisfying quasi-experimental design, because of its demonstration of all of the effects in all of the comparison groups. With awareness of the possible misinterpretations, it becomes a design well worth undertaking where better control is not possible. Having stressed its serious weaknesses, now let us examine and stress the relative strengths.

Like all quasi-experiments, this one gains strength through the consistency of the internal replications of the experiment. To make this consistency apparent, the main effects of occasions and of groups should be removed by expressing each cell as a deviation from the row (group) and column (time) means: $M_{gt} - M_{g.} - M_{.t} + M_{..}$ Then rearrange the data with treatments (*X*s) as column heads. Let us assume that the resulting picture is one of gratifying consistency, with the same treatment strongest in all four groups, etc. What are the chances of this being no true effect of treatments, but instead an interaction of groups and occasions? We can note that most possible interactions of groups and occasions would reduce or becloud the manifest effect of *X*. An interaction that imitates a main effect of *X* would be an unlikely one, and one that becomes more unlikely in larger Latin squares.

One would be most attracted to this design when one had scheduling control over a very few naturally aggregated groups, such as classrooms, but could not subdivide these natural groups into randomly equivalent subgroups for either presentation of *X* or for testing. For this situation, if pretesting is feasible, Design 10 is also available; it also involves a possible confounding of the effects of *X* with interactions of selection and occasions. This possibility is judged to be less likely in the counterbalanced design, because all comparisons are demonstrated in each group and hence several matched interactions would be required to imitate the experimental effect.

Whereas in the other designs the special responsiveness of just one of the groups to an extraneous event (history) or to practice (maturation) might simulate an effect of X_1, in the counterbalanced design such coincident effects would have to occur on separate occasions in each of the groups in turn. This assumes, of course, that we would not interpret a main effect of *X* as meaningful if inspection of the cells showed that a statistically significant main effect was primarily the result of a very strong effect in but one of the groups. For further discussion of this matter, see the reports of Wilk and Kempthorne (1957), Lubin (1961), and Stanley (1955).

12. The Separate-Sample Pretest-Posttest Design

For large populations, such as cities, factories, schools, and military units, it may often happen that although one cannot randomly segregate subgroups for differential experimental treatments, one can exercise something like full experimental control over the *when* and *to whom* of the O, employing random assignment procedures. Such control makes possible Design 12:

$$\begin{array}{lll} R & O & (X) \\ R & & X \quad O \end{array}$$

In this diagram, rows represent randomly equivalent subgroups, the parenthetical X standing for a presentation of X irrelevant to the argument. One sample is measured prior to the X, an equivalent one subsequent to X. The design is not inherently a strong one, as is indicated by its row in Table 2. Nevertheless, it may frequently be all that is feasible, and is often well worth doing. It has been used in social science experiments which remain the best studies extant on their topics (e.g., Star & Hughes, 1950). While it has been called the "simulated before-and-after design" (Selltiz, Jahoda, Deutsch, & Cook, 1959, p. 116), it is well to note its superiority over the ordinary before-and-after design, Design 2, through its control of both the main effect of testing and the interaction of testing with X. The main weakness of the design is its failure to control for history. Thus in the study of the Cincinnati publicity campaign for the United Nations and UNESCO (Star & Hughes, 1950), extraneous events on the international scene probably accounted for the observed decrease in optimism about getting along with Russia.

It is in the spirit of this chapter to encourage "patched-up" designs, in which features are added to control specific factors, more or less one at a time (in contrast with the neater "true" experiments, in which a single control group controls for all of the threats to internal validity). Repeating Design 12 in different settings at different times, as in Design 12*a* (see Table 2, p. 210), controls for history, in that if the same effect is repeatedly found, the likelihood of its being a product of coincidental historical events becomes less likely. But consistent secular historical trends or seasonal cycles still remain uncontrolled rival explanations. By replicating the effect under other settings, one can reduce the possibility that the observed effect is specific to the single population initially selected. However, if the setting of research permits Design 12*a*, it will also permit Design 13, which would in general be preferred.

Maturation, or the effect of the respondents' growing older, is unlikely to be invoked as a rival explanation, even in a public opinion survey study extending over months. But, in the sample survey setting, or even in some college classrooms, the samples are large enough and ages heterogeneous enough so that subsamples of the pretest group differing in maturation (age, number of semesters in college, etc.) can be compared. Maturation, and the probably more threatening possibility of secular and seasonal trends, can also be controlled by a design such as 12*b* which adds an additional earlier pretest group, moving the design closer to the time-series design, although without the repeated testing. For populations such as psychotherapy applicants, in which healing or spontaneous remission might take place, the assumptions of linearity implicitly involved in this control might not be plausible. It is more likely that the maturational trend will be negatively accelerated, hence will make the O_1—O_2 maturational gain larger than that for O_2—O_3, and thus work against the inter pretation that X has had an effect.

Instrumentation represents a hazard in this design when employed in the sample survey setting. If the same interviewers are employed in the pretest and in the posttest, it usually happens that many were doing their first interviewing on the pretest and are more experienced, or perhaps more cynical, on the posttest. If the interviewers differ on each wave and are few, differences in interviewer idiosyncrasies are confounded with the experimental variable. If the interviewers are aware of the hypothesis, and whether or not the X has been delivered, then interviewer expectations may create differences, as Stanton and Baker (1942) and Smith and Hyman (1950) have shown experimentally. Ideally, one would use equivalent random samples of different interviewers on each wave, and keep the interviewers in ignorance of the experiment. In addition, the recruitment of interviewers may show differences on a seasonal basis, for instance, because more college students are available during summer months, etc. Refusal rates are probably lower and interview lengths longer in summer than in winter. For questionnaires which are self-administered in the classroom, such instrument error may be less likely, although test-taking orientations may shift in ways perhaps better classifiable as instrumentation than as effects of X upon O.

For pretests and posttests separated in time by several months, mortality can be a problem in Design 12. If both samples are selected at the same time (point R), as time elapses, more members of the selected sample can be expected to become inaccessible, and the more transient segments of the population to be lost, producing a population difference between the different interviewing periods. Differences between groups in the number of noncontacted persons serve as a warning of this possibility.

Perhaps for studies over long periods the pretest and posttest samples should be selected independently and at appropriately different times, although this, too, has a source of systematic bias resulting from possible changes in the residential pattern of the universe as a whole. In some settings, as in schools, records will make possible the elimination of the pretest scores of those who have become unavailable by the time of the posttest, thus making the pretest and posttest more comparable. To provide a contact making this correction possible in the sample survey, and to provide an additional confirmation of effect which mortality could not contaminate, the pretest group can be retested, as in Design 12*c,* where the O_1—O_2 difference should confirm the O_1—O_3 comparison. Such was the study by Duncan, et al. (1957) on the reduction in fallacious beliefs effected by an introductory course in psychology. (In this design, the retested group does not make possible the examination of the gains for persons of various initial scores because of the absence of a control group to control for regression.)

It is characteristic of this design that it moves the laboratory into the field situation to which the researcher wishes to generalize, testing the effects of X in its natural setting. In general, as indicated in Tables 1 and 2, Designs 12, 12*a,* 12*b,* and 12*c* are apt to be superior in external validity or generalizability to the "true" experiments of Designs 4, 5, and 6. These designs put so little demand upon the respondents for cooperation, for being at certain places at certain times, etc., that representative sampling from populations specified in advance can be employed.

In Designs 12 and 13 (and, to be sure, in some variants on Designs 4 and 6, where X and O are delivered through individual contacts, etc.) representative sampling is possible. The pluses in the selection -X interaction column are highly relative and could, in justice, be changed to question marks, since in general practice the units are not selected for their theoretical relevance, but often for reasons of cooperativeness and accessibility, which make them likely to be atypical of the universe to which one wants to generalize.

It was not to Cincinnati but rather to Americans in general, or to people in general, that Star and Hughes (1950) wanted to generalize, and there remains the possibility that the reaction to X in Cincinnati was atypical of these universes. But the degree of such accessibility bias is so much less than that found in the more demanding designs that a comparative plus seems justified.

13. The Separate-Sample Pretest-Posttest Control Group Design

It is expected that Design 12 will be used in those settings in which the *X*, if presented at all, must be presented to the group as a whole. If there are comparable (if not equivalent) groups from which *X* can be withheld, then a control group can be added to Design 12, creating Design 13:

```
R O (X)
R     X O
---------
R O
R       O
```

This design is quite similar to Design 10, except that the same specific persons are not retested and thus the possible interaction of testing and *X* is avoided. As with Design 10, the weakness of Design 13 for internal validity comes from the possibility of mistaking for an effect of *X* a specific local trend in the experimental group which is, in fact, unrelated. By increasing the number of the social units involved (schools, cities, factories, ships, etc.) and by assigning them in some number and with randomization to the experimental and control treatments, the one source of invalidity can be removed, and a true experiment, like Design 4 except for avoiding the retesting of specific individuals, can be achieved. This design can be designated 13*a*. Its diagraming (in Table 3) has been complicated by the two levels of equivalence (achieved by random assignment) which are involved. At the level of respondents, there is within each social unit the equivalence of the separate pretest and posttest samples, indicated by the point of assignment *R*. Among the several social units receiving either treatment, there is no such equivalence, this lack being indicated by the dashed line. The *R′* designates the equation of the experimental group and the control group by the random assignment of these numerous social units to one or another treatment.

As can be seen by the row for 13*a* in Table 3, this design receives a perfect score for both internal and external validity, the latter on grounds already discussed for Design 12 with further strength on the selection-*X* interaction problem because of the representation of numerous social units, in contrast with the use of a single one. As far as is known, this excellent but expensive design has not been used.

14. The Multiple Time-Series Design

In studies of major administrative change by time-series data, the researcher would be wise to seek out a similar institution not undergoing the *X*, from which to collect a similar "control" time series (ideally with *X* assigned randomly):

```
O O O OXO O O O
---------------
O O O O  O O O O
```

This design contains within it (in the *O*s bracketing the *X*) Design 10, the Nonequivalent Control Group Design, but gains in certainty of interpretation from the multiple measures plotted, as the experimental effect is in a sense twice demonstrated, once against the control and once against the pre-*X* values in its own series, as in Design 7. In addition, the selection-maturation interaction is controlled to the extent that, if the experimental group showed in general a greater rate of gain, it would show up in the pre-*X* *O*s. In Tables 2 and 3 this additional gain is poorly represented, but appears in the final internal validity column, which is headed "Interaction of Selection and Maturation." Because maturation is controlled for both experimental and control series, by the logic discussed in the first presentation of the Time-Series Design 7 above, the difference in the selection of the groups operating in conjunction with maturation, instrumentation, or regression, can hardly account for an apparent effect. An interaction of the se-

TABLE 3
SOURCES OF INVALIDITY FOR QUASI-EXPERIMENTAL DESIGNS 13 THROUGH 16

	Sources of Invalidity											
	Internal								External			
	History	Maturation	Testing	Instrumentation	Regression	Selection	Mortality	Interaction of Selection and Maturation, etc.	Interaction of Testing and X	Interaction of Selection and X	Reactive Arrangements	Multiple-X Interference
Quasi-Experimental Designs Continued:												
13. Separate-Sample Pretest-Posttest Control Group Design *R O (X)* *R X O* *R O* *R O*	+	+	+	+	+	+	+	−	+	+	+	
13*a*. *R′* { *R O (X)* / *R X O* / *R O (X)* / *R X O* / *R O (X)* / *R X O* } *R′* { *R O* / *R O* / *R O* / *R O* / *R O* / *R O* }	+	+	+	+	+	+	+	+	+	+	+	
14. Multiple Time-Series *O O OXO O O* *O O O O O O O*	+	+	+	+	+	+	+	+	−	−	?	
15. Institutional Cycle Design Class *A X* O_1 Class B_1 RO_2 *X* O_3 Class B_2 *R X* O_4 Class *C* O_5 *X* [a]Gen. Pop. Con. Cl. *B* O_6 [a]Gen. Pop. Con. Cl. *C* O_7												
$O_2 < O_1$, $O_5 < O_4$	+	−	+	+	?	−	?		+	?	+	
$O_2 < O_3$	−	−	−	?	?	+	+		−	?	+	
$O_2 < O_4$	−	−	+	?	?	+	?		+	?	?	
$O_6 = O_7$, $O_{2y} = O_{2o}$		+						−				
16. Regression Discontinuity	+	+	+	?	+	+	?	+	+	−	+	+

[a] General Population Controls for Class B, etc.

lection difference with history remains, however, a possibility.

As with the Time-Series Design 7, a minus has been entered in the external validity column for testing-X interaction, although as with Design 7, the design would often be used where the testing was nonreactive. The standard precaution about the possible specificity of a demonstrated effect of X to the population under study is also recorded in Table 3. As to the tests of significance, it is suggested that differences between the experimental and control series be analyzed as Design 7 data. These differences seem much more likely to be linear than raw time-series data.

In general, this is an excellent quasi-experimental design, perhaps the best of the more feasible designs. It has clear advantages over Designs 7 and 10, as noted immediately above and in the Design 10 presentation. The availability of repeated measurements makes the Multiple Time Series particularly appropriate to research in schools.

15. The Recurrent Institutional Cycle Design: A "Patched-Up" Design

Design 15 illustrates a strategy for field research in which one starts out with an inadequate design and then adds specific features to control for one or another of the recurrent sources of invalidity. The result is often an inelegant accumulation of precautionary checks, which lacks the intrinsic symmetry of the "true" experimental designs, but nonetheless approaches experimentation. As a part of this strategy, the experimenter must be alert to the rival interpretations (other than an effect of X) which the design leaves open and must look for analyses of the data, or feasible extensions of the data, which will rule these out. Another feature often characteristic of such designs is that the effect of X is demonstrated in several different manners. This is obviously an important feature where each specific comparison would be equivocal by itself.

The specific "patched-up" design under discussion is limited to a narrow set of questions and settings, and opportunistically exploits features of these settings. The basic insight involved can be noted by an examination of the second and third rows of Table 1, in which it can be seen that the patterns of plus and minus marks for Designs 2 and 3 are for the most part complementary, and that hence the right combination of these two inadequate arguments might have considerable strength. The design is appropriate to those situations in which a given aspect of an institutional process is, on some cyclical schedule, continually being presented to a new group of respondents. Such situations include schools, indoctrination procedures, apprenticeships, etc. If in these situations one is interested in evaluating the effects of such a global and complex X as an indoctrination program, then the Recurrent Institutional Cycle Design probably offers as near an answer as is available from the designs developed thus far.

The design was originally conceptualized in the context of an investigation of the effects of one year's officer and pilot training upon the attitudes toward superiors and subordinates and leadership functions of a group of Air Force cadets in the process of completing a 14-month training cycle (Campbell & McCormack, 1957). The restriction precluding a true experiment was the inability to control who would be exposed to the experimental variable. There was no possibility of dividing the entering class into two equated halves, one half of which would be sent through the scheduled year's program, and the other half sent back to civilian life. Even were such a true experiment feasible (and opportunistic exploitation of unpredicted budget cuts might have on several occasions made such experiments possible), the reactive effects of such experimental arrangements, the disruption in the lives of those accepted, screened, and brought to the air base and then sent home, would have made them far from an ideal control group. The difference between them and the experimental group receiving indoctrination would hardly have been an adequate base from which to generalize to the normal conditions of recruitment and training. There remained, however, the experimenter's control over the scheduling of the *when* and *to whom* of the observational procedures. This, plus the fact that the experimental variable was recurrent and was continually being presented to a new group of respondents, made possible some degree of experimental control. In that study two kinds of comparisons relevant to the effect of military experience on attitudes were available. Each was quite inadequate in terms of experimental control, but when both provided confirmatory evidence they were mutually supportive inasmuch as they both involved different weaknesses. The first involved comparisons among populations measured at the same time but varying in their length of service. The second involved measures of the same group of persons in their first week of military training and then again after some 13 months. In idealized form this design is as follows:

Class A	X	O_1		
	—	—	—	—
Class B		O_2	X	O_3

This design combines the "longitudinal" and "cross-sectional" approaches commonly employed in developmental research. In this it is assumed that the scheduling is such that at one and the same time a group which has been exposed to X and a group which is just about to be exposed to it can be measured; this comparison between O_1 and O_2 thus corresponds to the Static-Group Comparison, Design 3. Remeasuring the personnel of Class B one cycle later provides the One-Group Pretest-Posttest segment, Design 2. In Table 3, on page 226, the first two rows dealing with Design 15 show an analysis of these comparisons. The cross-sectional comparison of $O_1 > O_2$ provides differences which could not be explained by the effects of history or a test-retest effect. The differences obtained could, however, be due to differences in recruitment from year to year (as indicated by the minus opposite selection) or by the fact that the respondents were one year older (the minus for maturation). Where the testing is all done at the same time period, the confounded variable of instrumentation, or shifts in the nature of the measuring instrument, seem unlikely. In the typical comparison of the differences in attitudes of freshmen and sophomores, the effect of mortality is also a rival explanation: O_1 and O_2 might differ just because of the kind of people that have dropped out from Class A but are still represented in Class B. This weakness is avoidable if the responses are identified by individuals, and if the experimenter waits before analyzing his data until Class B has completed its exposure to X and then eliminates from O_2 all of those measures belonging to respondents who later failed to complete the training. The frequent absence of this procedure justifies the insertion of a question mark opposite the mortality variable. The regression column is filled with question marks to warn of the possibility of spurious effects if the measure which is being used in the experimental design is the one on which the acceptance and rejection of candidates for the training course was based. Under these circumstances consistent differences which should not be attributed to the effects of X would be anticipated. The pretest-posttest comparison involved in O_2 and O_3, if it provides the same type of difference as does the $O_2 - O_1$ comparison, rules out the rival hypotheses that the difference is due to a shift in the selection or recruitment between the two classes, and also rules out any possibility that mortality is the explanation. However, were the $O_2 - O_3$ comparison to be used alone, it would be vulnerable to the rival explanations of history and testing.

In a setting where the training period under examination is one year, the most expensive feature of the design is the scheduling of the two sets of measurements a year apart. Given the investment already made in this, it constitutes little additional expense to do more testing on the second occasion. With this in mind, one can expand the recurrent institutional design to the pattern shown in Table 3. Exercising the power to designate who gets measured and when, Class B has been broken into two equated samples, one measured both before and after exposure, and the other measured only after exposure as in O_4. This second group provides a comparison on carefully equated samples of an initial measure coming before and after, is more precise than the $O_1 - O_2$ comparison as far as selection is concerned, and is superior to the $O_2 - O_3$ comparison in avoiding test-retest effects. The effect of X is thus documented in three separate comparisons, $O_1 > O_2$, $O_2 < O_3$ and $O_2 < O_4$.

Note, however, that O_2 is involved in all of these three, and thus all might appear to be confirmatory just because of an eccentric performance of that particular set of measurements. The introduction of O_5, that is Class C, tested on the second testing occasion prior to being exposed to X, provides another pre-X measure to be compared with O_4 and O_1, etc., providing a needed redundancy. The splitting of Class B makes this O_4—O_5 comparison more clear-cut than would be an O_3—O_5 comparison. Note, however, that the splitting of a class into the tested and the nontested half often constitutes a "reactive arrangement." For this reason a question mark has been inserted for that factor in the $O_2 < O_4$ row in Table 3. Whether or not this is a reactive procedure depends upon the specific conditions. Where lots are drawn and one half of the class is asked to go to another room, the procedure is likely to be reactive (e.g., Duncan, et al., 1957; Solomon, 1949). Where, as in many military studies, the contacts have been made individually, a class can be split into equated halves without this conspicuousness. Where a course consists of a number of sections with separate schedules, there is the possibility of assigning these intact units to the pretest and no-pretest groups (e.g., Hovland, Lumsdaine, & Sheffield, 1949). For a single classroom, the strategy of passing out questionnaires or tests to everyone but varying the content so that a random half would get what would constitute the pretest and the other half get tested on some other instrument may serve to make the splitting of the class no more reactive than the testing of the whole class would be.

The design as represented through measurements O_1 to O_5 uniformly fails to control for maturation. The seriousness of this limitation will vary depending upon the subject material under investigation. If the experiment deals with the acquisition of a highly esoteric skill or competence, the rival hypothesis of maturation—that just growing older or more experienced in normal everyday societal ways would have produced this gain—may seem highly unlikely.

In the cited study of attitudes toward superiors and subordinates (Campbell & McCormack, 1957), however, the shift was such that it might very plausibly be explained in terms of an increased sophistication which a group of that age and from that particular type of background would have undergone through growing older or being away from home in almost any context. In such a situation a control for maturation seems very essential. For this reason O_6 and O_7 have been added to the design, to provide a cross-sectional test of a general maturation hypothesis made on the occasion of the second testing period. This would involve testing two groups of persons from the general population who differ only in age and whose ages were picked to coincide with those of Class B and Class C at the time of testing. To confirm the hypothesis of an effect of X, the groups O_6 and O_7 should turn out to be equal, or at least to show less discrepancy than do the comparisons spanning exposure to X. The selection of these general population controls would depend upon the specificity of the hypothesis. Considering our knowledge as to the ubiquitous importance of social class and educational considerations, these controls might be selected so as to match the institutional recruitment on social class and previous education. They might also be persons who are living away from home for the first time and who are of the typical age of induction, so that, in the illustration given, the O_6 group would have been away from home one year and the O_7 group just barely on the verge of leaving home. These general population age-mate controls would always be to some extent unsatisfactory and would represent the greatest cost item, since testing within an institutional framework is generally easier than selecting cases from a general population. It is for this reason that O_6 and O_7 have been scheduled with the second testing wave, for if no effect of X is shown in the first body of results (the comparison $O_1 > O_2$), then these expensive procedures would usually be unjustified (unless, for example, one had the hypothesis that the institutional X had suppressed a normal maturational process).

Another cross-sectional approach to the control of maturation may be available if there is heterogeneity in age (or years away from home, etc.) within the population entering the institutional cycle. This would be so in many situations; for example, in studying the effects of a single college course. In this case, the measures of O_2 could be subdivided into an older and younger group to examine whether or not these two subgroups (O_{2_o} and O_{2_y} in Table 3) differed as did O_1 and O_2 (although the ubiquitous negative correlation between age and ability *within* school grades, etc., introduces dangers here). Better than the general population age-mate control might be the comparison with another specific institution, as comparing Air Force inductees with first-year college students. If the comparison is to be made of this type, one reduces one's experimental variable to those features which the two types of institution *do not* have in common. In this case, the generally more efficient Designs 10 and 13 would probably be as feasible.

The formal requirements of this design would seem to be applicable even to such a problem as that of psychotherapy. This possibility reveals how difficult a proper check on the maturation variable is. No matter how the general population controls for a psychotherapy situation are selected, if they are not themselves applicants for psychotherapy they differ in important ways. Even if they are just as ill as a psychotherapy applicant, they almost certainly differ in their awareness of, beliefs about, and faith in psychotherapy. Such an ill but optimistic group might very well have recovery potentialities not typical of any matching group that we would be likely to obtain, and thus an interaction of selection and maturation could be misinterpreted as an effect of X.

For the study of developmental processes per se, the failure to control maturation is of course no weakness, since maturation is the focus of study. This combination of longitudinal and cross-sectional comparisons should be more systematically employed in developmental studies. The cross-sectional study by itself confounds maturation with selection and mortality. The longitudinal study confounds maturation with repeated testing and with history. It alone is probably no better than the cross-sectional, although its greater cost gives it higher prestige. The combination, perhaps with repeated cross-sectional comparisons at various times, seems ideal.

In the diagrams of Design 15 as presented, it is assumed that it will be feasible to present the posttest for one group at the same chronological time as the pretest for another. This is not always the case in situations where we might want to use this design. The following is probably a more accurate portrayal of the typical opportunity in the school situation:

Class A	X	O_1					
Class B_1			RO_2	X	O_3		
Class B_2			R	X	O_4		
Class C						O_5	X

Such a design lacks the clear-cut control on history in the $O_1 > O_2$ and the $O_4 > O_5$ comparisons because of the absence of simultaneity. However, the explanation in terms of history could hardly be employed if both comparisons show the effect, except by postulating quite a complicated series of coincidences.

Note that any general historical trend, such as we certainly do find with social attitudes, is not confounded with clear-cut experimental results. Such a trend would make O_2 intermediate between O_1 and O_3, while the hypothesis that X has an effect requires O_1 and O_3 to be equal, and O_2 to differ from both in the same direction. In general, with replication of the experiment on several occasions, the confound with history is unlikely to be a problem even in this version of the design. But, for institutional cycles of less than a calendar year, there may be the possibility of confounding with seasonal variations in attitudes, morale, optimism, intelligence, or what have you. If the X is a course given only in the fall semester, and if between September and January people generally increase in hostility and pessimism because of seasonal climatic factors, this recurrent seasonal trend is confounded with the effects of X in all of its manifestations. For such settings, Designs 10 and 13 are available and to be preferred.

If the cross-sectional and longitudinal comparisons indicate comparable effects of X, this could not be explained away as an interaction between maturation and the selection differences between the classes. However, because this control does not show up in the segmental presentations in Table 3, the column has been left blank. The ratings on external validity criteria, in general, follow the pattern of the previous designs containing the same fragments. The question marks in the "Interaction of Selection and X" column merely warn that the findings are limited to the institutional cycle under study. Since the X is so complex, the investigation is apt to be made for practical reasons rather than theoretical purposes, and for these practical purposes, it is probably to this one institution that one wants to generalize in this case.

16. Regression-Discontinuity Analysis

This is a design developed in a situation in which ex post facto designs were previously being used. While very limited in range of possible applications, its presentation here seems justified by the fact that those limited settings are mainly educational. It also seems justifiable as an illustration of the desirability of exploring in each specific situation all of the implications of a causal hypothesis, seeking novel outcroppings where the hypothesis might be exposed to test. The setting (Thistlethwaite & Campbell, 1960) is one in which awards are made to the most qualified applicants on the basis of a cutting score on a quantified composite of qualifications. The award might be a scholarship, admission to a university so sought out that all accepted enrolled, a year's study in Europe, etc. Subsequent to this event, applicants receiving and not receiving the award are measured on various Os representing later achievements, attitudes, etc. The question is then asked, Did the award make a difference? The problem of inference is sticky just because almost all of the qualities leading to eligibility for the award (except such factors as need and state of residence, if relevant) are qualities which would have led to higher performance on these subsequent Os. We are virtually certain in advance that the recipients would have scored higher on the Os than the nonrecipients even if the award had not been made.

Figure 4 presents the argument of the design. It illustrates the expected relation of pre-award ability to later achievement, plus the added results of the special educational or motivational opportunities resulting. Let us first consider a true experiment of a Design 6 sort, with which to contrast our quasi-experiment. This true experiment might be rationalized as a tie-breaking process, or as an experiment in extension of program, in which, for a narrow range of scores at or just below the cutting point, random assignment would create an award-winning experimental group and a nonwinning control group. These would presumably perform as the two circle-points at the cutting line in Fig. 4. For this narrow range of abilities, a true experiment would have been achieved. *Such experiments are feasible and should be done.*

The quasi-experimental Design 16 attempts to substitute for this true experiment by examining the regression line for a discontinuity at the cutting point which the causal hypothesis clearly implies. If the outcome were as diagramed, and if the circle-points in Fig. 4 represented extrapolations from the two halves of the regression line rather than a randomly split tie-breaking experiment, the evidence of effect would be quite compelling, almost as compelling as in the case of the true experiment.

Some of the tests of significance discussed for Design 7 are relevant here. Note that the hypothesis is clearly one of intercept difference rather than slope, and that the location of the step in the regression line must be right at the *X* point, no "lags" or "spreads" being consistent with the hypothesis. Thus parametric and nonparametric tests avoiding assumptions of linearity are appropriate. Note also that assumptions of linearity are usually more plausible for such regression data than for time series. (For certain types of data, such as percentages, a linearizing transformation may be needed.) This might make a t test for the difference between the two linearly extrapolated points appropriate. Perhaps the most efficient test would be a covariance analysis, in which the award-decision score would be the covariate of later achievement, and award and no-award would be the treatment.

Is such a design likely to be used? It certainly applies to a recurrent situation in which claims for the efficacy of *X* abound. Are such claims worth testing? One sacrifice required is that all of the ingredients going into the final decision be pooled into a composite index, and that a cutting point be cleanly applied. But certainly we are convinced by now that all of the qualities lead-

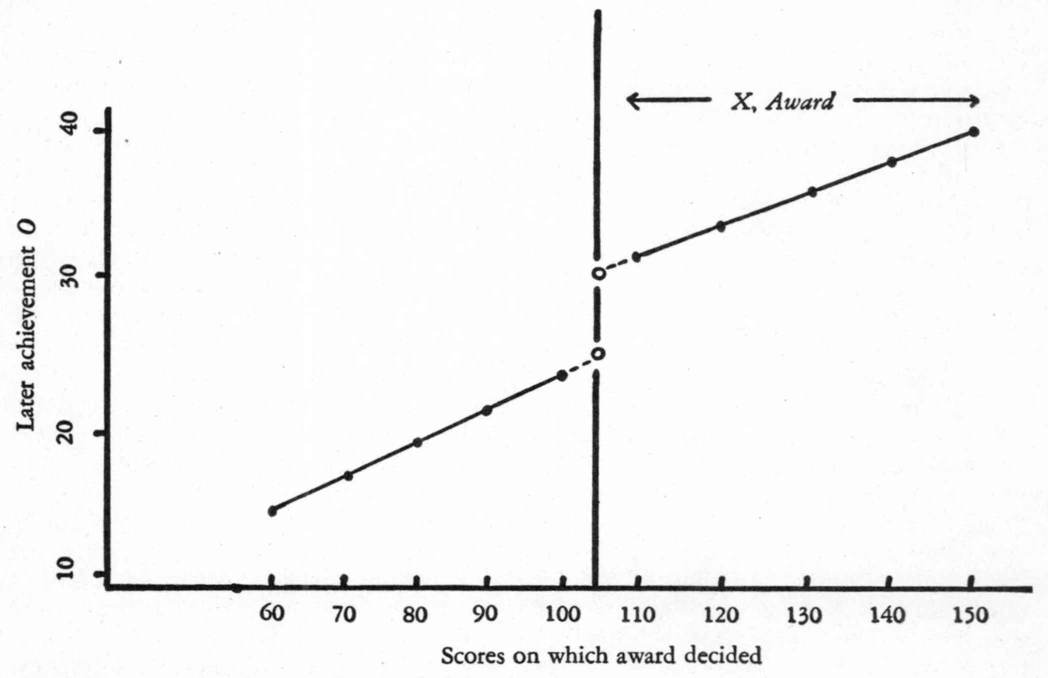

Fig. 4. Regression-Discontinuity Analysis.

ing to a decision—the appearance of the photograph, the class standing discounted by the high school's reputation, the college ties of the father, etc., can be put into such a composite, by ratings if by no more direct way. And we should likewise by now be convinced (Meehl, 1954) that a multiple correlational weighting formula for combining the ingredients (even using past committee decisions as a criterion) is usually better than a committee's case-by-case ponderings. Thus, we would have nothing to lose and much to gain for all purposes by quantifying award decisions of all kinds. If this were done, and if files were kept on awards and rejections, then years later follow-ups of effects could be made.

Perhaps a true parable is in order: A generous foundation interested in improving higher education once gave an Ivy League college half a million dollars to study the impact of the school upon its students. Ten years later, not a single research report remotely touching upon this purpose had appeared. Did the recipients or donors take the specifics of the formal proposal in any way seriously? Was the question in any way answerable? Designs 15 and 16 seem to offer the only possible approximations. But, of course, perhaps no scientist has any real curiosity about the effects of such a global X.

To go through the check-off in Table 3: Because of synchrony of experimental and control group, history and maturation seem controlled. Testing as a main effect is controlled in that both the experimental and control groups have received it. Instrumentation errors might well be a problem if the follow-up O was done under the auspices which made the award, in that gratitude for the award and resentment for not receiving the award might lead to differing expressions of attitude, differing degrees of exaggeration of one's own success in life, etc. This weakness would also be present in the tie-splitting true experiment. It could be controlled by having the follow-ups done by a separate agency. We believe, following the arguments above, that both regression and selection are controlled as far as their possible spurious contributions to inference are concerned, even though selection is biased and regression present—both have been controlled through representing them in detail, not through equation. Mortality would be a problem if the awarding agency conducted the follow-up measure, in that award recipients, alumni, etc., would probably cooperate much more readily than nonwinners. Note how the usually desirable wish of the researcher to achieve complete representation of the selected sample may be misleading here. If conducting the follow-up with a different letterhead would lead to a drop in cooperation from, say, 90 per cent to 50 per cent, an experimenter might be reluctant to make the shift because his goal is a 100 per cent representation of award winners. He is apt to forget that his true goal is interpretable data, that no data are interpretable in isolation, and that a comparable contrast group is essential to make use of his data on award winners. Both for this reason and because of the instrumentation problem, it might be scientifically better to have independent auspices and a 50 per cent return from both groups instead of a 90 per cent return from award winners and a 50 per cent return from the nonwinners. Again, the mortality problem would be the same for the tie-breaking true experiment. For both, the selection-maturation interaction threat to internal validity is controlled. For the quasi-experiment, it is controlled in that this interaction could not lawfully explain a distinct discontinuity in the regression line at X. The external validity threat of a testing-X interaction is controlled to the extent that the basic measurements used in the award decision are a part of the universe to which one wants to generalize.

Both the tie-breaking true experiment and the regression-discontinuity analysis are particularly subject to the external-validity limitation of selection-X interaction in that the effect has been demonstrated only for a very narrow band of talent, i.e., only for those at the cutting score. For the quasi-experiment, the possibilities of inference may seem broader, but note that the evils of the linear fit assumption are minimal when extrapolated but one point, as in the design as illustrated in Fig. 4. Broader generalizations involve the extrapolation of the below-*X* fit across the entire range of *X* values, and at each greater degree of extrapolation the number of plausible rival hypotheses becomes greater. Also, the extrapolated values of different types of curves fitted to the below-*X* values become more widely spread, etc.

CORRELATIONAL AND EX POST FACTO DESIGNS

One dimension of "*quasi*-ness" which has been increasing in the course of the last nine designs is the extent to which the *X* could be manipulated by the experimenter, i.e., could be intruded into the normal course of events. Certainly, the more this is so, the closer it is to true experimentation, as has been discussed in passing, particularly with regard to Designs 7 and 10. Designs 7, 10, 12, 13 (but not 13*a*), and 14 would be applicable both for naturally occurring *X*s and for *X*s deliberately introduced by the experimenter. The designs would be more suspect where the *X* was not under control, and some who might be willing to call the experimenter-controlled versions quasi-*experiments* might not be willing to apply this term to the uncontrolled *X*. We would not make an issue of this but would emphasize the value of data analyses of an experimental type for uncontrolled *X*s, as compared with the evaluational essays and misleading analyses too frequently used in these settings. Design 15 is, of course, completely limited to a naturally occurring *X*, and the designs of the present section (even if called data-analysis designs rather than quasi-experimental designs) are still more fully embedded in the natural setting. In this section, we will start again with the simple correlational analysis, then move to two designs of a fairly acceptable nature, and finally return to the ex post facto experiments, judged to be unsatisfactory at their very best.

Correlation and Causation

Design 3 is a correlational design of a very weak form, implying as it does the comparison of but two natural units, differing not only in the presence and absence of *X*, but also in innumerable other attributes. Each of these other attributes could create differences in the *O*s, and each therefore provides a plausible rival hypothesis to the hypothesis that *X* had an effect. We are left with a general rule that the differences between two natural objects are uninterpretable. Consider now this comparison expanded so that we have numerous independent natural instances of *X* and numerous ones of no-*X*, and concomitant differences in *O*. Insofar as the natural instances of *X* vary among each other in their other attributes, these other attributes become less plausible as rival hypotheses. Correlations of a fairly impressive nature may thus be established, such as that between heavy smoking and lung cancer. What is the status of such data as evidence of causation analogous to that provided by experiment?

A positive point may first be made. Such data are relevant to causal hypotheses inasmuch as they expose them to disconfirmation. If a zero correlation is obtained, the credibility of the hypothesis is lessened. If a high correlation occurs, the credibility of the hypothesis is strengthened in that it has survived a chance of disconfirmation. To put the matter another way, correlation does not necessarily indicate causation, but a causal law of the type producing mean differences in experiments does imply correlation. In any experiment where *X* has increased *O*, a positive biserial correlation between presence-absence of *X* and either posttest scores or gain scores will be found. The absence of such a correlation can rule out many simple, general, causal hypotheses, hypotheses as to main effects of *X*. In this sense, the relatively inexpensive correlational approach can provide a preliminary survey of hypotheses, and those which survive this can then be checked through the more expensive experimental manipulation. Katz, Maccoby, and Morse

(1951) have argued this and have provided a sequence in which the effects of leadership upon productivity were studied first correlationally, with a major hypothesis subsequently being checked experimentally (Morse & Reimer, 1956).

A perusal of research on teaching would soon convince one that the causal interpretation of correlational data is overdone rather than underdone, that plausible rival hypotheses are often overlooked, and that to establish the temporal antecedence-consequence of a causal relationship, observations extended in time, if not experimental intrusion of *X*, are essential. Where teacher's behavior and students' behavior are correlated, for example, our cultural stereotypes are such that we would almost never consider the possibility of the student's behavior causing the teacher's. Even when in a natural setting, an inherent temporal priority seems to be involved, selective retention processes can create a causality in the reverse direction. Consider, for example, possible findings that the superintendents with the better schools were better educated and that schools with frequent changes in superintendents had low morale. Almost inevitably we draw the implication that the educational level of superintendents and stable leadership *cause* better schools. The causal chain could be quite the reverse: better schools (for whatever reasons better) might cause well-educated men to stay on, while poorer schools might lead the better-educated men to be tempted away into other jobs. Likewise, better schools might well cause superintendents to stay in office longer. Still more ubiquitous than misleading reverse correlation is misleading third-variable correlation, in which the lawful determiners of who is exposed to *X* are of a nature which would also produce high *O* scores, even without the presence of *X*. To these instances we will return in the final section on the ex post facto design.

The true experiment differs from the correlational setting just because the process of randomization disrupts any lawful relationships between the character or antecedents of the students and their exposure to *X*. Where we have pretests and where clear-cut determination of who were exposed and who were not is available, then Designs 10 and 14 may be convincing even without the randomization. But for a design lacking a pretest (imitating Design 6) to occur naturally requires very special circumstances, which almost never happen. Even so, in keeping with our general emphasis upon the opportunistic exploitation of those settings which happen to provide interpretable data, one should keep his eyes open for them. Such settings will be those in which it seems plausible that exposure to *X* was lawless, arbitrary, uncorrelated with prior conditions. Ideally these arbitrary exposure decisions will also be numerous and mutually independent. Furthermore, they should be buttressed by whatever additional evidence is available, no matter how weak, as in the retrospective pretest discussed below. As Simon (1957, pp. 10–61) and Wold (1956) have in part argued, the causal interpretation of a simple or a partial correlation depends upon both the presence of a compatible plausible causal hypothesis and the absence of plausible rival hypotheses to explain the correlation upon other grounds.

One such correlational study is of such admirable opportunism as to deserve note here. Barch, Trumbo, and Nangle (1957) used the presence or absence of turn-signaling on the part of the car ahead as *X*, the presence or absence of turn-signaling by the following car as *O*, demonstrating a significant imitation, modeling, or conformity effect in agreement with many laboratory studies. Lacking any pretest, the interpretation is dependent upon the assumption of no relationship between the signaling tendencies of the two cars apart from the influence created by the behavior of the lead car. As published, the data seem compelling. Note, however, that any third variables which would affect the signaling frequency of both pairs of drivers in a similar fashion become plausible rival hypotheses. Thus if weather, degree of visibility, purpose of the driver as affected by time of day, presence of a parked police car, etc., have effects on both drivers, and if data are pooled across conditions heterogeneous in such third variables, the correlation can be explained without assuming any effect of the lead car's signaling per se. More interpretable as a "natural Design 6" is Brim's (1958) report on the effect of the sex of the sibling upon a child's personality in a two-child family. Sex determination may be nearly a perfect lottery. As far as is known, it is uncorrelated with the familial, social, and genetic determinants of personality. Third variable codetermination of sex of sibling and of a child's personality is at present not a plausible rival hypothesis to a causal interpretation of the interesting findings, nor is the reverse causation from personality of child to the sex of his sibling.

The Retrospective Pretest

In many military settings in wartime, it is plausible that the differing assignments among men of a common rank and specialty are made through chaotic processes, with negligible regard to special privileges, preferences, or capabilities. Therefore, a comparison of the attitudes of whites who happened to be assigned to racially mixed versus all-white combat infantry units can become of interest for its causal implications (Information and Education Division, 1947). We certainly should not turn our back on such data, but rather should seek supplementary data to rule out plausible rival hypotheses, keeping aware of the remaining sources of invalidity. In this instance, the "posttest" interview not only contained information about present attitudes toward Negroes (those in mixed companies being more favorable) but also asked for the recall of attitudes prior to the present assignment. These "retrospective pretests" showed no difference between the two groups, thus increasing the plausibility that prior to the assignment there had been no difference.

A similar analysis was important in a study by Deutsch and Collins (1951) comparing housing project occupants in integrated versus segregated units at a time of such housing shortage that people presumably took any available housing more or less regardless of their attitudes. Having only posttest measures, the differences they found might have been regarded as reflecting selection biases in initial attitudes. The interpretation that the integrated experience caused the more favorable attitudes was enhanced when a retrospective pretest showed no differences between the two types of housing groups in remembered prior attitudes. Given the autistic factors known to distort memory and interview reports, such data can never be crucial.

We long for the pretest entrance interview (and also for random assignment of tenants to treatments). Such studies are no doubt under way. But until supplanted by better data, the findings of Deutsch and Collins, including the retrospective pretest, are precious contributions to an experimentally oriented science in this difficult area.

The reader should be careful to note that the probable direction of memory bias is to distort the past attitudes into agreement with present ones, or into agreement with what the tenant has come to believe to be socially desirable attitudes. Thus memory bias seems more likely to disguise rather than masquerade as a significant effect of X in these instances.

If studies continue to be made comparing freshman and senior attitudes to show the impact of a college, the use of retrospective pretests to support the other comparisons would seem desirable as partial curbs to the rival hypotheses of history, selective mortality, and shifts in initial selection. (This is not to endorse any further repetition of such cross-sectional studies, when by now what we need are more longitudinal studies such as those of Newcomb, 1943, which provide repeated measures over the four-year period, supplemented by repeated cross-sectional surveys in the general manner of a four-year extension of Design 15. Let the necessarily hurried dissertations be done on other topics.)

Panel Studies

The simplest surveys represent observations at a single point in time, which often offer to the respondent the opportunity to classify himself as having been exposed to X or not exposed. To the correlations of exposure and posttest thus resulting there is contributed not only the common cause bias (in which the determinants of who gets X would also, even without X, cause high scores on O) but also a memory distortion with regard to X, further enhancing the spurious appearance of cause (Stouffer, 1950, p. 356). While such studies continue to support the causal inferences justifying advertising budgets (i.e., correlations between "Did you see the program?" and "Do you buy the product?"), they are trivial evidence of effect. They introduce a new factor threatening internal validity, i.e., biased misclassification of exposure to X, which we do not bother to enter into our tables.

In survey methodology, a great gain is made when the panel method, the repetition of interviews of the same persons, is introduced. At best, panel studies seem to provide the data for the weaker natural X version of Design 10 in instances in which exposure to some change agent, such as a motion picture or counseling contact, occurs between the two waves of interviews or questionnaires. The student in education must be warned, however, that within sociology this important methodological innovation is accompanied by a misleading analysis tradition. The "turnover table" (Glock, 1955), which is a cross-tabulation with percentages computed to subtotal bases, is extremely subject to the interpretative confounding of regression effects with causal hypotheses, as Campbell and Clayton (1961) pointed out. Even when analyzed in terms of pretest-posttest gains for an exposed versus a nonexposed group, a more subtle source of bias remains. In such a panel study, the exposure to the X (e.g., a widely seen antiprejudice motion picture) is ascertained in the second wave of the two-wave panel. The design is diagramed as follows:

$$\begin{pmatrix} O \\ O \end{pmatrix} \begin{pmatrix} X \quad O \\ ? \\ \quad\; O \end{pmatrix}$$

Two-wave Panel Design (unacceptable)

Here the spanning parentheses indicate occurrence of the O or X on the same interview; the question mark, ambiguity of classification into X and no-X groups. Unlike Design 10, the two-wave panel design is ambiguous as to who is in the control group and who in the experimental group. Like the worst studies of Design 10, the X is correlated with the pretest Os (in that the least prejudiced make most effort to go to the movie). But further than that, even if X had no true effect upon O, the correlation between X and the posttests would be higher than that between X and the pretest just because they occur on the same interview. It is a common experience in test and measurement research that any two items in the same questionnaire tend strongly to correlate more highly than do the same two if in separate questionnaires. Stockford and Bissell (1949) found adjacent items to correlate higher than nonadjacent ones even within the same instrument. Tests administered on the same day generally correlate higher than those administered on different days. In the panel study in question (Glock, 1955) the two interviews occurred some eight months apart. Sources of correlation enhancing those within one interview and lowering those across interviews include not only autonomous fluctuations in prejudice, but also differences in interviewers. The inevitable mistakes by the interviewer and misstatements by the interviewee in re-identifying former respondents result in some of the pretest-posttest pairs actually coming from different persons. The resulting higher X-posttest correlation implies that there will be less regression from X report to the posttest than to the pretest, and for this reason posttest differences in O will be greater than the pretest differences. This will result (if there has been no population gain whatsoever) in a pseudo gain for those self-classified as exposed and a pseudo loss for those self-classified as nonexposed. This outcome would usually be mistaken as confirming the hypothesis that X had an effect. (See Campbell & Clayton, 1961, for the details of this argument.)

To avoid this spurious source of higher correlation, the exposure to X might be ascertained independently of the interview, or in a separate intermediate wave of interviews. In the latter case, even if there were a biased memory for exposure, this should not artificially produce a higher X-posttest than X-pretest correlation. Such a design would be:

$$\left(\frac{O}{O}\right)\left(\frac{X}{?}\right)\left(\frac{O}{O}\right)$$

The Lazarsfeld Sixteenfold Table

Another ingenious quasi-experimental use of panel data, introduced by Lazarsfeld around 1948 in a mimeographed report entitled "The Mutual Effect of Statistical Variables," was initially intended to produce an index of the direction of causation (as well as of the strength of causation) existing between two variables. This analysis is currently known by the name of "the sixteenfold table" (e.g., Lipset, Lazarsfeld, Barton, & Linz, 1954, pp. 1160–1163), and is generally used to infer the relative strengths or depth of various attitudes rather than to infer the "direction of causation." It is this latter interest which makes it quasi-experimental.

Suppose that on a given occasion we can classify the behavior of 100 teachers as "warm" or "cold," and the behavior of their students as "responsive" or "unresponsive." Doing this, we discover a positive correlation: warm teachers have responsive classes. The question can now be asked, Does teacher warmth cause class responsiveness, or does a responsive class bring out warmth in teachers? While our cultural expectations prejudice us for the first interpretation, a very plausible case can be made for the second. (And, undoubtedly, reciprocal causation is involved.) A panel study would add relevant data by restudying the same variables upon a second occasion, with the same teachers and classes involved. (Two levels of measurement for two variables generate four response types for each occasion, or 4 × 4 possible response patterns for the two occasions, generating the sixteenfold table.) For illustrative purposes, assume this outcome:

FIRST OCCASION

Pupils	*Teachers* Cold	Warm
Responsive	20	30
Unresponsive	30	20

SECOND OCCASION

Pupils	*Teachers* Cold	Warm
Responsive	10	40
Unresponsive	40	10

The equivocality of ordinary correlational data and the ingenuity of Lazarsfeld's analysis become apparent if we note that among the shifts which would have made the transformation possible, these polar opposites exist:

TEACHER WARMTH
CAUSING PUPIL
RESPONSIVENESS

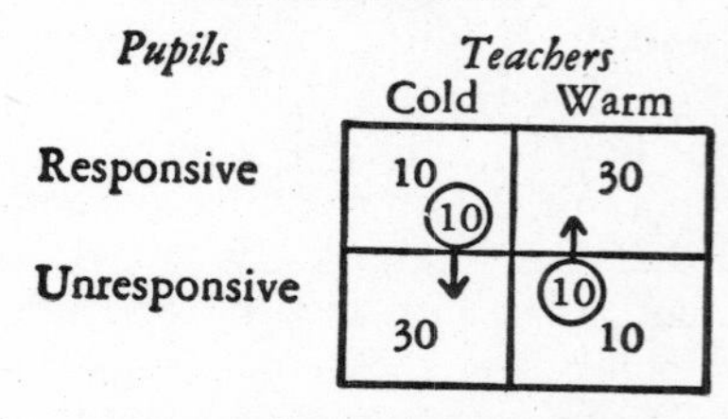

PUPIL RESPONSIVENESS
CAUSING TEACHER WARMTH

Pupils	*Teachers* Cold	Warm
Responsive	10 (10) →	30
Unresponsive	30	← (10) 10

Here we have considered only those changes increasing the correlation and have neglected the inevitable strays. Thus in this diagram, unlike Lazarsfeld's, we present only 8 of the 16 cells in his full sixteenfold table. We present only the four stable types (repeated in both top and bottom diagrams) and the four types of shifters whose shifting would increase the correlation (two in the top and two in the bottom). All four types of shifter could, of course, occur simultaneously, and any inference as to the direction of causation would be based upon a preponderance of one over the other. These diagrams represent the two most clear-cut outcomes possible. Were one of these to occur, then the examination of the character of the shifters, made possible by the panel type of data collection (impossible if different students and teachers were involved in each case), seems to add great plausibility to a one-directional causal inference. For those that shifted, the time dimension and the direction of change can be noted. If the first-shown case held, it would be implausible that students were changing teachers and highly plausible that teachers were changing students, at least for these 20 changing classrooms.

While the sociologists leave the analysis at the dichotomous level, these requirements can be restated more generally in terms of time-lagged correlations, in which the "effect" should correlate higher with a prior "cause" than with a subsequent "cause," i.e., $rx_1o_2 > rx_2o_1$. Taking the illustration of teachers causing pupils, we get:

Pupils Time 2	*Teachers Time 1* Cold	Warm
Responsive	10	40
Unresponsive	40	10

Pupils Time 1	*Teachers Time 2* Cold	Warm
Responsive	20	30
Unresponsive	30	20

In this instance the illustration seems a trivial restatement of the original tables because teachers did not change at all. This is, however, probably the best general form of the analysis. Note that while it is plausible, one probably should not use the argument $rx_1o_2 > rx_1o_1$ because of the many irrelevant sources of correlation occurring between data sets collected upon the same occasion which would inflate the rx_1o_1 value. It should be noted that the suggested $rx_1o_2 > rx_2o_1$ gives neither correlation an advantage in this respect.

What are the weaknesses of this design? Testing becomes a weakness in that repeated testing may quite generally result in higher correlations between correlated variables. The preliminary $rx_1o_1 < rx_2o_2$ may be explained away on these grounds. However, this could not easily explain away the $rx_1o_2 > rx_2o_1$ finding, unless an interaction or testing effect specific to but one of the variables were plausible.

Regression seems less of a problem for this design than for the two-wave panel study rejected above, since both *X* and *O* are assessed on both waves, and classifying in these terms is thus symmetrical. However, for the dichotomous Lazarsfeld-type analysis, regression does become a problem if the marginals of either variable are badly skewed (e.g., 10–90 splits rather than the 50–50 splits used in these illustrations). The analysis of correlations between continuous variables, using all cases, would not seem to encounter regression artifacts. Differential maturation upon the two variables, or differential effects of history, might be interaction effects threatening internal validity. With regard to external validity, the usual precautions hold, with particular emphasis upon the selection-*X* interaction in that the effect has been observed only for the subpopulation that shifts.

While in most teaching situations Designs 10 or 14 would be available and preferred for the type of problem used in our illustration, there are probably settings in which this analysis should be considered. For example,

Dr. Winfred F. Hill has suggested the application of the analysis to data on parent and child behavior as collected in longitudinal studies.[6]

When generalized to nondichotomous data, the name "Sixteenfold Table" becomes inappropriate; we recommend the title "Cross-Lagged Panel Correlation" for this analysis.

Ex Post Facto Analyses

The phrase "ex post facto experiment" has come to refer to efforts to simulate experimentation through a process of attempting in a Design 3 situation to accomplish a pre-*X* equation by a process of matching on pre-*X* attributes. The mode of analysis and name were first introduced by Chapin (Chapin & Queen, 1937). Subsequently this design has been treated extensively by Greenwood (1945) and Chapin (1947, 1955). While these citations come from sociology rather than education, and while we judge the analysis a misleading one, treatment in this Handbook seems appropriate. It represents one of the most extended efforts toward quasi-experimental design. The illustrations are frequently from education. The mode of thinking employed and the errors involved are recurrent in educational research also.

In one typical ex post facto study (Chapin, 1955, pp. 99–124) the *X* was high school education (particularly finishing high school) and the *O*s dealt with success and community adjustment ten years later, as judged from information obtained in individual interviews. The matching in this case was done from records retained in the high school files (although in similar, still weaker studies these pre-*X* facts are obtained in the post-*X* interviews). Initially the data showed those completing high school to have been more successful but also to have had higher marks in grammar school, higher parental occupations, younger ages, better neighborhoods, etc. Thus these antecedents might have caused both completion of high school and later success. Did the schooling have any additional effect over and above the head start provided by these background factors? Chapin's "solution" to this question was to examine subsets of students matched on all these background factors but differing in completion of high school. The addition of each matching factor reduced in turn the posttest discrepancy between the *X* and no-*X* groups, but when all matching was done, a significant difference remained. Chapin concluded, although cautiously, that education had an effect. An initial universe of 2,127 students shrank to 1,194 completed interviews on cases with adequate records. Matching then shrunk the usable cases to 46, i.e., 23 graduates and 23 nongraduates, less than 4 per cent of those interviewed. Chapin well argues that 46 comparable cases are better than 1,194 noncomparable ones on grounds similar to our emphasis upon the priority of internal validity over external validity. The tragedy is that his 46 cases are still not comparable, and furthermore, even within his faulty argument the shrinkage was unnecessary.

He has seriously *under*matched for two distinct reasons. His first source of undermatching is that matching is subject to differential regression, which would certainly produce in this case a final difference in the direction obtained (after the manner indicated by R. L. Thorndike, 1942, and discussed with regard to matching in Design 10, above). The direction of the pseudo effect of regression to group means after matching is certain in this case, because the differences in the matching factors for those successful versus unsuccessful are in the same direction for each factor as the differences between those completing versus those not completing high school. Every determinant of exposure to *X* is likewise, even without *X*, a determinant of *O*. All matching variables correlate with *X* and *O* in the same direction. While this might not be so of every variable in all ex post facto studies, it is the case in most if not all published examples. This error

[6] Personal communication.

and the reduction in number of cases are avoidable through the modern statistics which supplanted the matching-error in Design 10. The matching variables could all be used as covariates in a multiple-covariate analysis of covariance. It is our considered estimate that this analysis would remove the apparently significant effects in the specific studies which Chapin presents. (But see Lord, 1960, for his criticism of the analysis of covariance for such problems.) There is, however, a second and essentially uncorrectable source of undermatching in Chapin's setting. Greenwood (1945) refers to it as the fact of self-selection of exposure or nonexposure. Exposure is a lawful product of numerous antecedents. In the case of dropping out of high school before completion, we know that there are innumerable determinants beyond the six upon which matching was done. We can with great assurance surmise that most of these will have a similar effect upon later success, independently of their effect through X. This insures that there will be undermatching over and above the matching-regression effect. Even with the pre-X-predictor and O covariance analysis, a significant treatment effect is interpretable only when *all* of the jointly contributing matching variables have been included.

CONCLUDING REMARKS

Since a handbook chapter is already a condensed treatment, further condensation is apt to prove misleading. In this regard, a final word of caution is needed about the tendency to use the speciously convenient Tables 1, 2, and 3 for this purpose. These tables have added a degree of order to the chapter as a recurrent outline and have made it possible for the text to be less repetitious than it would otherwise have been. But the placing of specific pluses and minuses and question marks has been continually equivocal and usually an inadequate summary of the corresponding discussion. For any specific execution of a design, the check-off row would probably be different from the corresponding row in the table. Note, for example, that the tie-breaking case of Design 6 discussed incidentally in connection with quasi-experimental Design 16 has, according to that discussion, two question marks and one minus not appearing in the Design 6 row of Table 1. The tables are better used as an outline for a conscientious scrutiny of the specific details of an experiment while planning it. Similarly, this chapter is not intended to substitute a dogma of *the* 13 acceptable designs for an earlier dogma of *the* one or *the* two acceptable. Rather, it should encourage an open-minded and exploratory orientation to novel data-collection arrangements and a new scrutiny of some of the weaknesses that accompany routine utilizations of the traditional ones.

In conclusion, in this chapter we have discussed alternatives in the arrangement or design of experiments, with particular regard to the problems of control of extraneous variables and threats to validity. A distinction has been made between internal validity and external validity, or generalizability. Eight classes of threats to internal validity and four factors jeopardizing external validity have been employed to evaluate 16 experimental designs and some variations on them. Three of these designs have been classified as pre-experimental and have been employed primarily to illustrate the validity factors needing control. Three designs have been classified as "true" experimental designs. Ten designs have been classified as quasi-experiments lacking optimal control but worth undertaking where better designs are impossible. In interpreting the results of such experiments, the check list of validity factors becomes particularly important. Throughout, attention has been called to the possibility of creatively utilizing the idiosyncratic features of any specific research situation in designing unique tests of causal hypotheses.

REFERENCES

Allport, F. H. The influence of the group upon association and thought. *J. exp. Psychol.*, 1920, 3, 159–182.

Anastasi, Anne. *Differential psychology.* (3rd ed.) New York: Macmillan, 1958.

Anderson, N. H. Test of a model for opinion change. *J. abnorm. soc. Psychol.,* 1959, 59, 371–381.

Barch, A. M., Trumbo, D., & Nangle, J. Social setting and conformity to a legal requirement. *J. abnorm. soc. Psychol.,* 1957, 55, 396–398.

Boring, E. G. The nature and the history of experimental control. *Amer. J. Psychol.,* 1954, 67, 573–589.

Brim, O. G. Family structure and sex role learning by children: A further analysis of Helen Koch's data. *Sociometry,* 1958, 21, 1–16.

Brolyer, C. R., Thorndike, E. L., & Woodyard, Ella. A second study of mental discipline in high school studies. *J. educ. Psychol.,* 1927, 18, 377–404.

Brownlee, K. A. *Statistical theory and methodology in science and engineering.* New York: Wiley, 1960.

Brunswik, E. *Perception and the representative design of psychological experiments.* (2nd ed.) Berkeley: Univer. of California Press, 1956.

Campbell, D. T. Factors relevant to the validity of experiments in social settings. *Psychol. Bull.,* 1957, 54, 297–312.

Campbell, D. T. Methodological suggestions from a comparative psychology of knowledge processes. *Inquiry,* 1959, 2, 152–182.

Campbell, D. T. Recommendations for APA test standards regarding construct, trait, or discriminant validity. *Amer. Psychologist,* 1960, 15, 546–553.

Campbell, D. T. Quasi-experimental designs for use in natural social settings. In D. T. Campbell, *Experimenting, validating, knowing: Problems of method in the social sciences.* New York: McGraw-Hill, in preparation.

Campbell, D. T., & Clayton, K. N. Avoiding regression effects in panel studies of communication impact. *Stud. pub. Commun.,* 1961, No. 3, 99–118.

Campbell, D. T., & Fiske, D. W. Convergent and discriminant validation by the multitrait-multimethod matrix. *Psychol. Bull.,* 1959, 56, 81–105.

Campbell, D. T., & McCormack, Thelma H. Military experience and attitudes toward authority. *Amer. J. Sociol.,* 1957, 62, 482–490.

Cane, V. R., & Heim, A. W. The effects of repeated testing: III. Further experiments and general conclusions. *Quart. J. exp. Psychol.,* 1950, 2, 182–195.

Cantor, G. N. A note on a methodological error commonly committed in medical and psychological research. *Amer. J. ment. Defic.,* 1956, 61, 17–18.

Chapin, F. S. *Experimental designs in sociological research.* New York: Harper, 1947; (Rev. ed., 1955).

Chapin, F. S., & Queen, S. A. *Research memorandum on social work in the depression.* New York: Social Science Research Council, Bull. 39, 1937.

Chernoff, H., & Moses, L. E. *Elementary decision theory.* New York: Wiley, 1959.

Cochran, W. G., & Cox, Gertrude M. *Experimental designs.* (2nd ed.) New York: Wiley, 1957.

Collier, R. M. The effect of propaganda upon attitude following a critical examination of the propaganda itself. *J. soc. Psychol.,* 1944, 20, 3–17.

Collier, R. O., Jr. Three types of randomization in a two-factor experiment. Minneapolis: Author, 1960. (Dittoed)

Cornfield, J., & Tukey, J. W. Average values of mean squares in factorials. *Ann. math. Statist.,* 1956, 27, 907–949.

Cox, D. R. Some systematic experimental designs. *Biometrika,* 1951, 38, 312–323.

Cox, D. R. The use of a concomitant variable in selecting an experimental design. *Biometrika,* 1957, 44, 150–158.

Cox, D. R. *Planning of experiments.* New York: Wiley, 1958.

Crook, M. N. The constancy of neuroticism scores and self-judgments of constancy. *J. Psychol.,* 1937, 4, 27–34.

Deutsch, M., & Collins, Mary E. *Interracial housing: A psychological evaluation of a social experiment.* Minneapolis: Univer. of Minnesota Press, 1951.

Duncan, C. P., O'Brien, R. B., Murray, D. C., Davis, L., & Gilliland, A. R. Some information about a test of psychological misconceptions. *J. gen. Psychol.,* 1957, 56, 257–260.

Ebbinghaus, H. *Memory.* Trans. by H. A. Ruger and C. E. Bussenius. New York: Teachers Coll., Columbia Univer., 1913. (Original, *Über das Gedächtnis,* Leipzig, 1885.)

Edwards, A. L. *Experimental design in psy-*

chological research. (Rev. ed.) New York: Rinehart, 1960.

Farmer, E., Brooks, R. C., & Chambers, E. G. *A comparison of different shift systems in the glass trade.* Rep. 24, Medical Research Council, Industrial Fatigue Research Board. London: His Majesty's Stationery Office, 1923.

Feldt, L. S. A comparison of the precision of three experimental designs employing a concomitant variable. *Psychometrika,* 1958, 23, 335–353.

Ferguson, G. A. *Statistical analysis in psychology and education.* New York: McGraw-Hill, 1959.

Fisher, R. A. *Statistical methods for research workers.* (1st ed.) London: Oliver & Boyd, 1925.

Fisher, R. A. *The design of experiments.* (1st ed.) London: Oliver & Boyd, 1935.

Fisher, R. A. The arrangement of field experiments. *J. Min. Agriculture,* 1926, 33, 503–513; also in R. A. Fisher, *Contributions to mathematical statistics.* New York: Wiley, 1950.

Glickman, S. E. Perseverative neural processes and consolidation of the memory trace. *Psychol. Bull.,* 1961, 58, 218–233.

Glock, C. Y. Some applications of the panel method to the study of social change. In P. F. Lazarsfeld & M. Rosenberg (Eds.), *The language of social research.* Glencoe, Ill.: Free Press, 1955. Pp. 242–249.

Glock, C. Y. The effects of re-interviewing in panel research. 1958. Multilith of a chapter to appear in P. F. Lazarsfeld (Ed.), *The study of short run social change,* in preparation.

Good, C. V., & Scates, D. E. *Methods of research.* New York: Appleton-Century-Crofts, 1954.

Grant, D. A. Analysis-of-variance tests in the analysis and comparison of curves. *Psychol. Bull.,* 1956, 53, 141–154.

Green, B. F., & Tukey, J. W. Complex analyses of variance: General problems. *Psychometrika,* 1960, 25, 127–152.

Greenwood, E. *Experimental sociology: A study in method.* New York: King's Crown Press, 1945.

Guetzkow, H., Kelly, E. L., & McKeachie, W. J. An experimental comparison of recitation, discussion, and tutorial methods in college teaching. *J. educ. Psychol.,* 1954, 45, 193–207.

Hammond, K. R. Representative vs. systematic design in clinical psychology. *Psychol. Bull.,* 1954, 51, 150–159.

Hanson, N. R. *Patterns of discovery.* Cambridge, Eng.: Univer. Press, 1958.

Hovland, C. I., Janis, I. L., & Kelley, H. H. *Communication and persuasion.* New Haven, Conn.: Yale Univer. Press, 1953.

Hovland, C. I., Lumsdaine, A. A., & Sheffield, F. D. *Experiments on mass communication.* Princeton, N.J.: Princeton Univer. Press, 1949.

Information and Education Division, U. S. War Department. Opinions about Negro infantry platoons in white companies of seven divisions. In T. M. Newcomb & E. L. Hartley (Eds.), *Readings in social psychology.* New York: Holt, 1947. Pp. 542–546.

Johnson, P. O. *Statistical methods in research.* New York: Prentice-Hall, 1949.

Johnson, P. O., & Jackson, R. W. B. *Modern statistical methods: Descriptive and inductive.* Chicago: Rand McNally, 1959.

Jost, A. Die Assoziationsfestigkeit in ihrer Abhängigkeit von der Verteilung der Wiederholungen. *Z. Psychol. Physiol. Sinnesorgane,* 1897, 14, 436–472.

Kaiser, H. F. Directional statistical decisions. *Psychol. Rev.,* 1960, 67, 160–167.

Katz, D., Maccoby, N., & Morse, Nancy C. *Productivity, supervision, and morale in an office situation.* Ann Arbor: Survey Research Center, Univer. of Michigan, 1951.

Kempthorne, O. *The design and analysis of experiments.* New York: Wiley, 1952.

Kempthorne, O. The randomization theory of statistical inference. *J. Amer. Statist. Ass.,* 1955, 50, 946–967; 1956, 51, 651.

Kempthorne, O. The design and analysis of experiments, with some reference to educational research. In R. O. Collier & S. M. Elam (Eds.), *Research design and analysis: The second annual Phi Delta Kappa symposium on educational research.* Bloomington, Ind.: Phi Delta Kappa, 1961. Pp. 97–133.

Kendall, M. G., & Buckland, W. R. *A dictionary of statistical terms.* London: Oliver & Boyd, 1957.

Kennedy, J. L., & Uphoff, H. F. Experiments on the nature of extra-sensory perception. III. The recording error criticisms of extra-chance scores. *J. Parapsychol.,* 1939, 3, 226–245.

Kerr, W. A. Experiments on the effect of music on factory production. *Appl. Psychol. Monogr.*, 1945, No. 5.

Lana, R. E. Pretest-treatment interaction effects in attitudinal studies. *Psychol. Bull.*, 1959, 56, 293–300. (a)

Lana, R. E. A further investigation of the pretest-treatment interaction effect. *J. appl. Psychol.*, 1959, 43, 421–422. (b)

Lana, R. E., & King, D. J. Learning factors as determiners of pretest sensitization. *J. appl. Psychol.*, 1960, 44, 189–191.

Lindquist, E. F. *Statistical analysis in educational research.* Boston: Houghton Mifflin, 1940.

Lindquist, E. F. *Design and analysis of experiments in psychology and education.* Boston: Houghton Mifflin, 1953.

Lipset, S. M., Lazarsfeld, P. F., Barton, A. H., & Linz, J. The psychology of voting: An analysis of political behavior. In G. Lindzey (Ed.), *Handbook of social psychology.* Cambridge, Mass.: Addison-Wesley, 1954. Pp. 1124–1175.

Lord, F. M. The measurement of growth. *Educ. psychol. Measmt,* 1956, 16, 421–437.

Lord, F. M. Further problems in the measurement of growth. *Educ. psychol. Measmt,* 1958, 18, 437–451.

Lord, F. M. Large-sample covariance analysis when the control variable is fallible. *J. Amer. Statist. Ass.*, 1960, 55, 307–321.

Lubin, A. The interpretation of significant interaction. *Educ. psychol. Measmt,* 1961, 21, 807–817.

Maxwell, A. E. *Experimental design in psychology and the medical sciences.* London: Methuen, 1958.

McCall, W. A. *How to experiment in education.* New York: Macmillan, 1923.

McNemar, Q. A critical examination of the University of Iowa studies of environmental influences upon the I.Q. *Psychol. Bull.*, 1940, 37, 63–92.

McNemar, Q. *Psychological statistics.* (3rd ed.) New York: Wiley, 1962.

McNemar, Q. On growth measurement. *Educ. psychol. Measmt,* 1958, 18, 47–55.

Meehl, P. E. *Clinical versus statistical prediction.* Minneapolis: Univer. of Minnesota Press, 1954.

Monroe, W. S. General methods: Classroom experimentation. In G. M. Whipple (Ed.), *Yearb. nat. Soc. Stud. Educ.*, 1938, 37, Part II, 319–327.

Mood, A. F. *Introduction to the theory of statistics.* New York: McGraw-Hill, 1950.

Moore, H. T. The comparative influence of majority and expert opinion. *Amer. J. Psychol.*, 1921, 32, 16–20.

Morse, Nancy C., & Reimer, E. The experimental change of a major organizational variable. *J. abnorm. soc. Psychol.*, 1956, 52, 120–129.

Myers, J. L. On the interaction of two scaled variables. *Psychol. Bull.*, 1959, 56, 384–391.

Newcomb, T. M. *Personality and social change.* New York: Dryden, 1943.

Neyman, J. Indeterminism in science and new demands on statisticians. *J. Amer. Statist. Ass.*, 1960, 55, 625–639.

Nunnally, J. The place of statistics in psychology. *Educ. psychol. Measmt,* 1960, 20, 641–650.

Page, E. B. Teacher comments and student performance: A seventy-four classroom experiment in school motivation. *J. educ. Psychol.*, 1958, 49, 173–181.

Pearson, H. C. Experimental studies in the teaching of spelling. *Teachers Coll. Rec.*, 1912, 13, 37–66.

Peters, C. C., & Van Voorhis, W. R. *Statistical procedures and their mathematical bases.* New York: McGraw-Hill, 1940.

Piers, Ellen V. Effects of instruction on teacher attitudes: Extended control-group design. Unpublished doctoral dissertation, George Peabody Coll., 1954. Abstract in *Bull. Maritime Psychol. Ass.*, 1955 (Spring), 53–56.

Popper, K. R. *The logic of scientific discovery.* New York: Basic Books, 1959.

Rankin, R. E., & Campbell, D. T. Galvanic skin response to Negro and white experimenters. *J. abnorm. soc. Psychol.*, 1955, 51, 30–33.

Reed, J. C. Some effects of short term training in reading under conditions of controlled motivation. *J. educ. Psychol.*, 1956, 47, 257–264.

Rogers, C. R., & Dymond, Rosalind F. *Psychotherapy and personality change.* Chicago: Univer. of Chicago Press, 1954.

Rosenthal, R. Research on experimenter bias. Paper read at Amer. Psychol. Ass., Cincinnati, Sept., 1959.

Roy, S. N., & Gnanadesikan, R. Some contributions to ANOVA in one or more dimensions: I and II. *Ann. math. Statist.,* 1959, 30, 304–317, 318–340.

Rozeboom, W. W. The fallacy of the null-hypothesis significance test. *Psychol. Bull.,* 1960, 57, 416–428.

Rulon, P. J. Problems of regression. *Harvard educ. Rev.,* 1941, 11, 213–223.

Sanford, F. H., & Hemphill, J. K. An evaluation of a brief course in psychology at the U.S. Naval Academy. *Educ. psychol. Measmt,* 1952, 12, 194–216.

Scheffé, H. Alternative models for the analysis of variance. *Ann. math. Statist.,* 1956, 27, 251–271.

Selltiz, Claire, Jahoda, Marie, Deutsch, M., & Cook, S. W. *Research methods in social relations.* (Rev. ed.) New York: Holt-Dryden, 1959.

Siegel, Alberta E., & Siegel, S. Reference groups, membership groups, and attitude change. *J. abnorm. soc. Psychol.,* 1957, 55, 360–364.

Simon, H. A. *Models of man.* New York: Wiley, 1957.

Smith, H. L., & Hyman, H. The biasing effect of interviewer expectations on survey results. *Publ. opin. Quart.,* 1950, 14, 491–506.

Sobol, M. G. Panel mortality and panel bias. *J. Amer. Statist. Ass.,* 1959, 54, 52–68.

Solomon, R. L. An extension of control group design. *Psychol. Bull.,* 1949, 46, 137–150.

Sorokin, P. A. An experimental study of efficiency of work under various specified conditions. *Amer. J. Sociol.,* 1930, 35, 765–782.

Stanley, J. C. Statistical analysis of scores from counterbalanced tests. *J. exp. Educ.,* 1955, 23, 187–207.

Stanley, J. C. Fixed, random, and mixed models in the analysis of variance as special cases of finite model III. *Psychol. Rep.,* 1956, 2, 369.

Stanley, J. C. Controlled experimentation in the classroom. *J. exp. Educ.,* 1957, 25, 195–201. (a)

Stanley, J. C. Research methods: Experimental design. *Rev. educ. Res.,* 1957, 27, 449–459. (b)

Stanley, J. C. Interactions of organisms with experimental variables as a key to the integration of organismic and variable-manipulating research. In Edith M. Huddleston (Ed.), *Yearb. Nat. Counc. Measmt used in Educ.,* 1960, 7–13.

Stanley, J. C. Analysis of a doubly nested design. *Educ. psychol. Measmt,* 1961, 21, 831–837. (a)

Stanley, J. C. Studying status vs. manipulating variables. In R. O. Collier & S. M. Elam (Eds.), *Research design and analysis: The second Phi Delta Kappa symposium on educational research.* Bloomington, Ind.: Phi Delta Kappa, 1961. Pp. 173–208. (b)

Stanley, J. C. Analysis of unreplicated three-way classifications, with applications to rater bias and trait independence. *Psychometrika,* 1961, 26, 205–220. (c)

Stanley, J. C. Analysis-of-variance principles applied to the grading of essay tests. *J. exp. Educ.,* 1962, 30, 279–283.

Stanley, J. C., & Beeman, Ellen Y. Interaction of major field of study with kind of test. *Psychol. Rep.,* 1956, 2, 333–336.

Stanley, J. C., & Beeman, Ellen Y. Restricted generalization, bias, and loss of power that may result from matching groups. *Psychol. Newsltr,* 1958, 9, 88–102.

Stanley, J. C., & Wiley, D. E. *Development and analysis of experimental designs for ratings.* Madison, Wis.: Authors, 1962.

Stanton, F., & Baker, K. H. Interviewer-bias and the recall of incompletely learned materials. *Sociometry,* 1942, 5, 123–134.

Star, Shirley A., & Hughes, Helen M. Report on an educational campaign: The Cincinnati plan for the United Nations. *Amer. J. Sociol.,* 1950, 55, 389–400.

Stockford, L., & Bissell, H. W. Factors involved in establishing a merit-rating scale. *Personnel,* 1949, 26, 94–116.

Stouffer, S. A. (Ed.) *The American soldier.* Princeton, N.J.: Princeton Univer. Press, 1949. Vols. I & II.

Stouffer, S. A. Some observations on study design. *Amer. J. Sociol.,* 1950, 55, 355–361.

Thistlethwaite, D. L., & Campbell, D. T. Regression-discontinuity analysis: An alternative to the ex post facto experiment. *J. educ. Psychol.,* 1960, 51, 309–317.

Thorndike, E. L., & Woodworth, R. S. The influence of improvement in one mental function upon the efficiency of other functions. *Psychol. Rev.,* 1901, 8, 247–261, 384–395, 553–564.

Thorndike, E. L., McCall, W. A., & Chapman, J. C. Ventilation in relation to mental work. *Teach. Coll. Contr. Educ.,* 1916, No. 78.

Thorndike, R. L. Regression fallacies in the matched groups experiment. *Psychometrika,* 1942, 7, 85–102.

Underwood, B. J. *Experimental psychology.* New York: Appleton-Century-Crofts, 1949.

Underwood, B. J. An analysis of the methodology used to investigate thinking behavior. Paper read at New York Univer. Conf. on Human Problem Solving. April, 1954. (See also C. I. Hovland & H. H. Kendler, The New York University Conference on Human Problem Solving. *Amer. Psychologist,* 1955, 10, 64–68.)

Underwood, B. J. Interference and forgetting. *Psychol. Rev.,* 1957, 64, 49–60. (a)

Underwood, B. J. *Psychological research.* New York: Appleton-Century-Crofts, 1957. (b)

Underwood, B. J., & Richardson, J. Studies of distributed practice. XVIII. The influence of meaningfulness and intralist similarity of serial nonsense lists. *J. exp. Psychol.,* 1958, 56, 213–219.

Watson, R. I. *Psychology of the child.* New York: Wiley, 1959.

Wilk, M. B., & Kempthorne, O. Fixed, mixed, and random models. *J. Amer. Statist. Ass.,* 1955, 50, 1144–1167; Corrigenda, *J. Amer. Statist. Ass.,* 1956, 51, 652.

Wilk, M. B., & Kempthorne, O. Some aspects of the analysis of factorial experiments in a completely randomized design. *Ann. math. Statist.,* 1956, 27, 950–985.

Wilk, M. B., & Kempthorne, O. Non-additivities in a Latin square design. *J. Amer. Statist. Ass.,* 1957, 52, 218–236.

Windle, C. Test-retest effect on personality questionnaires. *Educ. psychol. Measmt,* 1954, 14, 617–633.

Winer, B. J. *Statistical principles in experimental design.* New York: McGraw-Hill, 1962.

Wold, H. Causal inference from observational data. A review of ends and means. *J. Royal Statist. Soc.,* Sec. A., 1956, 119, 28–61.

Wyatt, S., Fraser, J. A., & Stock, F. G. L. *Fan ventilation in a humid weaving shed.* Rept. 37, Medical Research Council, Industrial Fatigue Research Board. London: His Majesty's Stationery Office, 1926.

Zeisel, H. *Say it with figures.* New York: Harper, 1947.

Some Supplementary References

Blalock, H. M. *Causal inferences in nonexperimental research.* Chapel Hill: Univer. of North Carolina Press, 1964.

Box, G. E. P. Bayesian approaches to some bothersome problems in data analysis. In J. C. Stanley (Ed.), *Improving experimental design and statistical analysis.* Chicago: Rand McNally, 1967.

Box, G. E. P., & Tiao, G. C. A change in level of a non-stationary time series. *Biometrika,* 1965, 52, 181–192.

Campbell, D. T. From description to experimentation: Interpreting trends as quasi-experiments. In C. W. Harris (Ed.), *Problems in measuring change.* Madison: Univer. of Wisconsin Press, 1963. Pp. 212–242.

Campbell, D. T. Administrative experimentation, institutional records, and nonreactive measures. In J. C. Stanley (Ed.), *Improving experimental design and statistical analysis.* Chicago: Rand McNally, 1967.

Glass, G. V. Evaluating testing, maturation, and treatment effects in a pretest posttest quasi-experimental design. *Amer. educ. res. J.,* 1965, 2, 83–87.

Pelz, D. C., & Andrews, F. M. Detecting causal priorities in panel study data. *Amer. sociol. Rev.,* 1964, 29, 836–848.

Stanley, J. C. Quasi-experimentation. *Sch. Rev.,* 1965, 73, 197–205.

Stanley, J. C. A common class of pseudo-experiments. *Amer. educ. res. J.,* 1966, 3, 79–87.

Stanley, J. C. The influence of Fisher's *The design of experiments* on educational research thirty years later. *Amer. educ. res. J.,* 1966, 3, 223–229.

Stanley J. C. Rice as a pioneer educational researcher. *J. educ. Measmt,* June, 1966, 3, 135–139.

Webb, E. J., Campbell, D. T., Schwartz, R. D., & Sechrest, L. *Unobtrusive measures: Nonreactive research in the social sciences* Chicago: Rand McNally, 1966.

Name Index

Subject Index

PRINTED IN U.S.A